Ben Stacy Jerrik (Ed.)

Brodhead, Kentucky

Ben Stacy Jerrik (Ed.)

Brodhead, Kentucky

Rockcastle County, Kentucky, United States, 2000 United States Census

Part Press

Imprint

Publisher:
Part Press is a trademark of
International Book Market Service Ltd., 17 Rue Meldrum, Beau Bassin, 1713-01 Mauritius
Email: info@bookmarketservice.com
Website: www.bookmarketservice.com

Published in 2013

Printed in: U.S.A., U.K., Germany. This book was not produced in Mauritius.

ISBN: 978-620-1-60870-2

Contents

Articles

References

Montalchez

Montalchez	
Country Switzerland **Canton** Neuchâtel **District** Boudry	
Population	244 (Dec 2011)[1]
- Density	38 /km^2 (99 /sq mi)
Area	km^2 (sq mi)[2]
Elevation	656 m (ft)
Postal code	2027
SFOS number	6411
Surrounded by	Fresens, Gorgier, Provence (VD), Saint-Aubin-Sauges
Website	Montalchez.ch [3] Profile [4] (French), SFSO statistics [5]

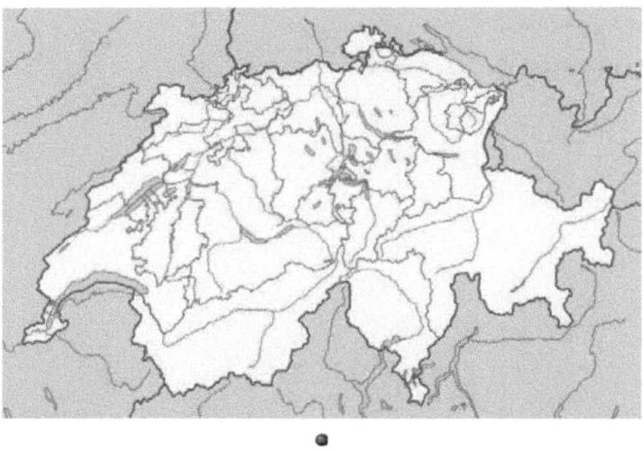

Montalchez

Montalchez is a municipality in the district of Boudry in the canton of Neuchâtel in Switzerland.

History

Montalchez is first mentioned in 1340 as *villa de Montallichie*.[6]

Geography

Montalchez has an area, as of 2009, of 6.4 square kilometers (sq mi). Of this area, 3.72 km^2 (sq mi) or 57.9% is used for agricultural purposes, while 2.41 km^2 (sq mi) or 37.5% is forested. Of the rest of the land, 0.27 km^2 (sq mi) or 4.2% is settled (buildings or roads) and 0.01 km^2 (acres) or 0.2% is unproductive land.[7]

Of the built up area, housing and buildings made up 2.3% and transportation infrastructure made up 1.6%. Out of the forested land, 33.8% of the total land area is heavily forested and 3.7% is covered with orchards or small clusters of trees. Of the agricultural land, 17.9% is used for growing crops and 22.1% is pastures, while 1.2% is used for orchards or vine crops and 16.7% is used for alpine pastures.[7]

The municipality is located in the Boudry district. It consists of the village of Montalchez and the hamlet of Erperens.

Coat of arms

The blazon of the municipal coat of arms is *Per fess Argent and Gules, a Rose counterchanged.*[8]

Demographics

Montalchez has a population (as of December 2011) of 244.[1] As of 2008, 4.4% of the population are resident foreign nationals.[9] Over the last 10 years (2000–2010) the population has changed at a rate of 37.6%. It has changed at a rate of 27.1% due to migration and at a rate of 10% due to births and deaths.[10]

Most of the population (as of 2000) speaks French (179 or 97.3%) as their first language, German is the second most common (3 or 1.6%) and English is the third (1 or 0.5%).[11]

As of 2008, the population was 52.5% male and 47.5% female. The population was made up of 122 Swiss men (51.7% of the population) and 2 (0.8%) non-Swiss men. There were 105 Swiss women (44.5%) and 7 (3.0%) non-Swiss women.[12] Of the population in the municipality, 75 or about 40.8% were born in Montalchez and lived there in 2000. There were 43 or 23.4% who were born in the same canton, while 43 or 23.4% were born somewhere else in Switzerland, and 15 or 8.2% were born outside of Switzerland.[11]

As of 2000, children and teenagers (0–19 years old) make up 21.2% of the population, while adults (20–64 years old) make up 60.3% and seniors (over 64 years old) make up 18.5%.[10]

As of 2000, there were 78 people who were single and never married in the municipality. There were 87 married individuals, 11 widows or widowers and 8 individuals who are divorced.[11]

As of 2000, there were 72 private households in the municipality, and an average of 2.5 persons per household.[10] There were 17 households that consist of only one person and 5 households with five or more people. In 2000, a total of 70 apartments (63.1% of the total) were permanently occupied, while 28 apartments (25.2%) were seasonally occupied and 13 apartments (11.7%) were empty.[13] As of 2009, the construction rate of new housing units was 21.4 new units per 1000 residents.[10]

The historical population is given in the following chart:[6] [14]

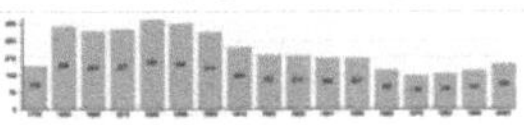

Politics

In the 2007 federal election the most popular party was the SVP which received 29.15% of the vote. The next three most popular parties were the LPS Party (18.39%), the SP (15.7%) and the FDP (13.23%). In the federal election, a total of 91 votes were cast, and the voter turnout was 54.5%.[15]

Economy

As of 2010, Montalchez had an unemployment rate of 2%. As of 2008, there were 34 people employed in the primary economic sector and about 11 businesses involved in this sector. No one was employed in the secondary sector. 10 people were employed in the tertiary sector, with 6 businesses in this sector.[10] There were 104 residents of the municipality who were employed in some capacity, of which females made up 39.4% of the workforce.

In 2008 the total number of full-time equivalent jobs was 35. The number of jobs in the primary sector was 26, all of which were in agriculture. There were no jobs in the secondary sector. The number of jobs in the tertiary sector was 9, of which 6 were in wholesale or retail sales or the repair of motor vehicles, 2 were in a hotel or restaurant and 1 was in the information industry.[16]

In 2000, there were 71 workers who commuted away from the municipality.[17] Of the working population, 5.8% used public transportation to get to work, and 67.3% used a private car.[10]

Religion

From the 2000 census, 18 or 9.8% were Roman Catholic, while 128 or 69.6% belonged to the Swiss Reformed Church. Of the rest of the population, there were 24 individuals (or about 13.04% of the population) who belonged to another Christian church. 20 (or about 10.87% of the population) belonged to no church, are agnostic or atheist, and 6 individuals (or about 3.26% of the population) did not answer the question.[11]

Education

In Montalchez about 69 or (37.5%) of the population have completed non-mandatory upper secondary education, and 14 or (7.6%) have completed additional higher education (either university or a *Fachhochschule*). Of the 14 who completed tertiary schooling, 57.1% were Swiss men, 14.3% were Swiss women.[11]

In the canton of Neuchâtel most municipalities provide two years of non-mandatory kindergarten, followed by five years of mandatory primary education. The next four years of mandatory secondary education is provided at thirteen larger secondary schools, which many students travel out of their home municipality to attend.[18] Both the kindergarten and the primary school are combined with Fresens. During the 2010-11 school year, there was one kindergarten class with a total of 23 students between the municipalities. In the same year, there was one primary class with a total of 16 students.[19]

As of 2000, there were 20 students from Montalchez who attended schools outside the municipality.[17]

References

[1] Swiss Federal Statistics Office – STAT-TAB (http://www.pxweb.bfs.admin.ch/Dialog/varval.asp?ma=px-d-01-2A01&ti=Ständige+und+Nichtständige+Wohnbevölkerung+nach+Region,+Geschlecht,+Nationalität+und++Alter&path=../Database/German_01 - Bevölkerung/01.2 - Bevölkerungsstand und -bewegung/&lang=1&prod=01&openChild=true&secprod=2) *Ständige und Nichtständige Wohnbevölkerung nach Region, Geschlecht, Nationalität und Alter* (German) accessed 3 October 2012

[2] Arealstatistik Standard - Gemeindedaten nach 4 Hauptbereichen (http://www.bfs.admin.ch/bfs/portal/de/index/themen/02/03/blank/data/gemeindedaten.html)

[3] http://Montalchez.ch/

[4] http://www.ne.ch/neat/documents/statistiques/Fiches_4829/wportraits_Files/NE6411.pdf

[5] http://www.bfs.admin.ch/bfs/portal/en/index/regionen/02/key.html

[6] *Montalchez* in German (http://www.hls-dhs-dss.ch/textes/d/D2833.php), French (http://www.hls-dhs-dss.ch/textes/f/F2833.php) and Italian (http://www.hls-dhs-dss.ch/textes/i/I2833.php) in the online *Historical Dictionary of Switzerland*.

[7] Swiss Federal Statistical Office-Land Use Statistics (http://www.bfs.admin.ch/bfs/portal/de/index/themen/02/03/blank/data/gemeindedaten.html) 2009 data (German) accessed 25 March 2010

[8] Flags of the World.com (http://www.crwflags.com/fotw/flags/ch-ne011.html) accessed 20-October-2011

[9] Swiss Federal Statistical Office - Superweb database - Gemeinde Statistics 1981-2008 (http://www.bfs.admin.ch/bfs/portal/de/index/infothek/onlinedb/superweb/login.html) (German) accessed 19 June 2010

[10] Swiss Federal Statistical Office (http://www.bfs.admin.ch/bfs/portal/en/index/regionen/02/key.html) accessed 20-October-2011

[11] STAT-TAB Datenwürfel für Thema 40.3 - 2000 (http://www.pxweb.bfs.admin.ch/Database/German_40 - EidgenÃ¶ssische VolkszÃ¤hlung/40.3 - 2000/40.3 - 2000.asp?lang=1&prod=40&secprod=3&openChild=true) (German) accessed 2 February 2011

[12] Canton of Neuchatel Statistics (http://www.ne.ch/neat/site/jsp/rubrique/rubrique.jsp?StyleType=bleu&DocId=40383), *République et canton de Neuchâtel - Recensement annuel de la population* (German) accessed 13 October 2011

[13] Swiss Federal Statistical Office STAT-TAB - Datenwürfel für Thema 09.2 - Gebäude und Wohnungen (http://www.pxweb.bfs.admin.ch/Database/German_09 - Bau- und Wohnungswesen/09.2 - GebÃ¤ude und Wohnungen/09.2 - GebÃ¤ude und Wohnungen.asp?lang=1&prod=09&secprod=2&openChild=true) (German) accessed 28 January 2011

[14] Swiss Federal Statistical Office STAT-TAB Bevölkerungsentwicklung nach Region, 1850-2000 (http://www.pxweb.bfs.admin.ch/Database/German_40 - EidgenÃ¶ssische VolkszÃ¤hlung/40.4 - 1850-2000/40.4 - 1850-2000.asp?lang=1&prod=40&secprod=4&openChild=true) (German) accessed 29 January 2011

[15] Swiss Federal Statistical Office, *Nationalratswahlen 2007: Stärke der Parteien und Wahlbeteiligung, nach Gemeinden/Bezirk/Canton* (http://www.bfs.admin.ch/bfs/portal/de/index/themen/17/02/blank/data/04/03.html) (German) accessed 28 May 2010

[16] Swiss Federal Statistical Office STAT-TAB Betriebszählung: Arbeitsstätten nach Gemeinde und NOGA 2008 (Abschnitte), Sektoren 1-3 (http://www.pxweb.bfs.admin.ch/Database/German_06 - Industrie und Dienstleistungen/06.2 - Unternehmen/06.2 - Unternehmen.asp?lang=1&prod=06&secprod=2&openChild=true) (German) accessed 28 January 2011

[17] Swiss Federal Statistical Office - Statweb (http://www.media-stat.admin.ch/stat/pendler/pop.php) (German) accessed 24 June 2010

[18] EDK/CDIP/IDES (2010). Kantonale Schulstrukturen in der Schweiz und im Fürstentum Liechtenstein / Structures Scolaires Cantonales en Suisse et Dans la Principauté du Liechtenstein (http://edudoc.ch/record/35128/files/Schulsystem_alle.pdf) (Report). . Retrieved 24 June 2010.

[19] Statistical Department of the Canton of Neuchâtel (http://www.ne.ch/neat/site/jsp/rubrique/rubrique.jsp?CatId=9939) Mémento de l'année scolaire 2010/2011 (French) accessed 17 October 2011

Communes_of_Switzerland

Communes (German: *Gemeinden* / French: *communes* / Italian: *comuni* / Romansh: *vischnancas*), also known as municipalities, are the smallest government division in Switzerland, numbering 2,596 (as of February 2010).[1] While many have a population of a few hundred citizens, the largest cities such as Zürich or Geneva also have the legal status of municipalities. The area of the municipalities varies between 0.28 km² (Ponte Tresa, Ticino) and 430 km² (Glarus Süd, Glarus).

Map of Switzerland showing municipal boundaries (2012)

Each canton defines the responsibilities of its constituent communes. These may include providing local government services such as education, medical and social services, public transportation, and tax collection. The degree of centralization varies from one canton to another.

Communes are generally governed by a council (sometimes called Municipality) headed by a mayor as executive and the town meeting as legislature. Most cantons leave the option to larger municipalities to opt for a city parliament. In some cantons, foreign persons that have dwelled for a certain time in Switzerland are also allowed to participate in the municipal politics.

Swiss citizenship is based on the citizenship of a municipality. Every Swiss is citizen of one or several municipalities (known as the place of origin, *lieu d'origine*, or *Heimatort*). As on the cantonal and federal level, citizens enjoy their political rights, including the direct democratic ones, in their municipality.

Communes are financed through direct taxes (such as the income tax), with rates varying more or less within a framework set by the canton. As among the cantons, there is a tax transfer among the communes to balance various levels of tax income.

Many municipalities are having difficulties maintaining the civil services they need to perform the duties they are required to do. In an effort to reduce expenses, some municipalities are combining together (through mergers or the creation of special-purpose districts). This restructuring is generally encouraged by the cantonal governments and the rate of these unions is increasing.

"Cities" (*villes* or *Städte*) are the municipalities with more than 10'000 inhabitants, or smaller places which had medieval town rights. There is no specific designation for smaller communities such as "village" or "town".

Population	Number of municipalities in 2004 (%)[2]
>20,000	30 (1.1%)
10,000–19,999	89 (3.2%)
5000–9999	180 (6.6%)
1000–4999	1025 (37.4%)
500–999	555 (20.3%)
<500	861 (31.4%)
Total	2740 (100%)

Communes are numbered by the Swiss Federal Office for Statistics (see *Community Identification Number#Switzerland*). One or more postal codes (PLZ/NPA) can by assigned to a municipality or shared with other municipalities.

See also: List of cities in Switzerland

Lists of communes by canton

- Aargau: *Einwohnergemeinden*
- Appenzell I.R.: *Bezirke*
- Appenzell A.R.: *Einwohnergemeinden*
- Basel-City: *Einwohnergemeinden*
- Basel-Country: *Einwohnergemeinden*
- Bern: *Einwohnergemeinden* or *communes municipales*
- Fribourg: *communes* or *Gemeinden*
- Geneva: *communes*
- Glarus: *Ortsgemeinden*
- Grisons: *politische Gemeinden, Vischnancas politicas* or *comuni politici*
- Jura: *communes*
- Lucerne: *Einwohnergemeinden*
- Neuchâtel: *communes*
- Nidwalden: *Gemeinden*
- Obwalden: *Einwohnergemeinden*
- Schaffhausen: *Einwohnergemeinden*
- Schwyz: *Gemeinden*
- Solothurn: *Einwohnergemeinden*
- St. Gallen: *Politische Gemeinden*
- Thurgau: *politische Gemeinden*
- Ticino: *comuni politici*
- Uri: *Einwohnergemeinde*
- Valais: *communes municipales* or *Einwohnergemeinden*
- Vaud: *communes*
- Zug: *Einwohnergemeinde*
- Zürich: *Politische Gemeinden*

See cantons of Switzerland for the number of municipalities per canton.

List of communes by population

Largest municipalities

City	C.	Pop.
Zürich	ZH	376,815
Geneva	GE	185,726
Basel	BS	167,365
Bern	BE	128,041
Lausanne	VD	129,273
Winterthur	ZH	100,000
St. Gallen	SG	74,538
Lucerne	LU	57,890
Lugano	TI	56,719

Biel/Bienne	BE	50,852
Thun	BE	42,319
Köniz	BE	38,098
La Chaux-de-Fonds	NE	37,023
Schaffhausen	SH	33,527
Fribourg	FR	33,008
Chur	GR	34,915
Neuchâtel	NE	32,389
Vernier	GE	32,038
Uster	ZH	28,770
Sion	VS	27,697
Emmen	LU	27,274

Smallest municipalities (As of 2002)

Rank	Municipality	C.	Pop.
1.	Portein	GR	22
2.	Corippo	TI	23
3.	Mulegns	GR	25
4.	Bister	VS	26
5.	Martisberg	VS	28
6.	Goumoens-le-Jux	VD	30
6.	Largario	TI	30
8.	Clugin	GR	35
9.	Cauco	GR	36
9.	Monible	BE	36
11.	St. Martin	GR	37
12.	Gresso	TI	38
12.	Montfavergier	JU	38
14.	Vaugondry	VD	39
15.	Arrissoules	VD	40
15.	Romairon	VD	40
15.	Selma	GR	40
18.	Ausserbinn	VS	41
18.	Linescio	TI	41
18.	Mauraz	VD	41
18.	Steinhaus	VS	41

nb. Corippo has now taken place as smallest city/municipality in Switzerland with only 17 inhabitants. (2004 Census)

History

The beginnings of the modern municipality system date back to the Helvetic Republic. Under the Old Swiss Confederacy, citizenship was granted by each town and village to only residents. These citizens enjoyed access to community property and in some cases additional protection under the law. Additionally, the urban towns and the rural villages had differing rights and laws. The creation of a uniform Swiss citizenship, which applied equally for citizens of the old towns and their tenants and servants, led to conflict. The wealthier villagers and urban citizens held rights rights to forests, common land and other municipal property which they did not want to share with the "new citizens", who were generally poor. The compromise solution, which was written into the municipal laws of the Helvetic Republic, is still valid today. Two politically separate but often geographically similar organizations were created. The first, the so-called municipality, was a political community formed by election and its voting body consists of all resident citizens. However, the community land and property remained with the former local citizens who were gathered together into the *Bürgergemeinde*. During the Mediation era (1803–1814), and especially during the Restoration era (1814–1830), many of the gains toward uniform citizenship were lost. Many political municipalities were abolished and limits were placed on the exercise of political rights for everyone except the members of the *Bürgergemeinde*. In the Regeneration era (1830–1848), the liberal revolutions of the common people helped to restore some rights again in a few cantons. In other cantons, the *Bürgergemeinden* were able to maintain power as political communities. In the city of Zurich it wasn't until the Municipal Act of 1866 that the political municipality came back into existence.[3]

The relationship between the political municipality and the *Bürgergemeinde* was often dominated by the latter's ownership of community property. Often the administration and profit from the property were totally held by the *Bürgergemeinden*, leaving the political municipality dependent on the *Bürgergemeinde* for money and use of the property. It wasn't until the political municipality acquired rights over property that served the public (such as schools, fire stations, etc.) and taxes, that they obtained full independence. For example, in the city of Bern, it wasn't until after the property division of 1852 that the political municipality had the right to levy taxes.[3]

It wasn't until the Federal Constitution of 1874 that all Swiss citizens were granted equal political rights on local and Federal levels. This revised constitution finally removed all the political voting and electoral body rights from the *Bürgergemeinde*. In the cities, the percentage of members in the *Bürgergemeinde* in the population was reduced as a result of increasing emigration to the cities. This led to the *Bürgergemeinde* losing its former importance to a large extent. However, the *Bürgergemeinde* has remained, and it includes all individuals who are citizens of the *Bürgergemeinde*, usually by having inherited the *Bürgerrecht* (citizenship), regardless of where they were born or where they may currently live. Instead of the place of birth, Swiss legal documents, e.g. passports, contain the *Bürgerort* (place of citizenship). The *Bürgergemeinde* also often holds and administers the common property in the village for the members of the community.[3]

Other types of communes

In addition to the political communes or municipalities, a number of other communes exist in Switzerland. These include:

- Bürgergemeinde (also: Burgergemeinde, Ortsgemeinde, Ortsbürgergemeinde, Tagwen, bourgeoisie, commune bourgeoise, vischnanca burgaisa), a statutory corporation that includes everyone who is a citizen of a commune and has the *Heimatrecht* (home right) in that commune regardless of where they may currently live. Until the 19th Century this *Heimatrecht* included rights to use the commons, which were administered by the Bürgergemeinde. Modernly, some Bürgergemeinden may still control common property, but the *Heimatrecht* and associated *Heimatort* is used just as place of birth in other countries.
- Gemischte Gemeinde (mixed communes), found in the Canton of Jura and portions of the Canton of Bern, a combination of a Bürgergemeinde and a political commune.

- Korporationsgemeinde, a legally recognized cooperative in Central Switzerland that controls some land and is responsible for its members support.
- Kirchgemeinde, a parish for members of a large church (generally Roman Catholic or Swiss Reformed. There may be two or more *Kirchgemeinden* in a single municipality.
- Schulgemeinde, similar to a School district.
- Bäuert, in the Berner Oberland or Graubünden) a small farming community. It is a type of agricultural cooperative with shared equipment and land.[4]
- Degagna, in the Leventina valley in the Canton of Ticino. It manages shared pastures, fields and woods as well as maintaining roads that cross the common land.[5]

See also

- List of municipalities of Switzerland
- List of twin towns and sister cities in Switzerland
- Former municipalities of Switzerland

Notes and references

[2] Official list of Swiss municipalities, p. 17

[3] *Bürgergemeinde* in German (http://www.hls-dhs-dss.ch/textes/d/D26443.php), French (http://www.hls-dhs-dss.ch/textes/f/F26443.php) and Italian (http://www.hls-dhs-dss.ch/textes/i/I26443.php) in the online *Historical Dictionary of Switzerland.*

[4] *Bäuert* in German (http://www.hls-dhs-dss.ch/textes/d/D10405.php), French (http://www.hls-dhs-dss.ch/textes/f/F10405.php) and Italian (http://www.hls-dhs-dss.ch/textes/i/I10405.php) in the online *Historical Dictionary of Switzerland.*

[5] *Degagna* in German (http://www.hls-dhs-dss.ch/textes/d/D35272.php), French (http://www.hls-dhs-dss.ch/textes/f/F35272.php) and Italian (http://www.hls-dhs-dss.ch/textes/i/I35272.php) in the online *Historical Dictionary of Switzerland.*

External links

- Amtliches Gemeindeverzeichnis der Schweiz/Liste officielle des communes de la Suisse/Elenco ufficiale dei Comuni della Svizzera, 2006 (http://www.bfs.admin.ch/bfs/portal/de/index/news/publikationen.html?publicationID=2321). Neuchâtel, 2006. ISBN 3-303-00334-3.
- *Nachbarschaft* in German (http://www.hls-dhs-dss.ch/textes/d/D16403.php), French (http://www.hls-dhs-dss.ch/textes/f/F16403.php) and Italian (http://www.hls-dhs-dss.ch/textes/i/I16403.php) in the online *Historical Dictionary of Switzerland.*

Boudry_District

Boudry District is one of the six districts of the canton of Neuchâtel, Switzerland. It has a population of 38845 (as of 31 December 2011).[1] The district capital is the town of Boudry.

Municipalities

The district consists of the following municipalities:

Coat of arms	Municipality	Population (31 December 2011)[1]	Area km^2[2]
	Bevaix	3836	10.78
	Boudry	4952	16.77
	Brot-Dessous	95	4.95
	Corcelles-Cormondrèche	4657	4.86
	Cortaillod	4576	3.69
	Fresens	222	1.58
	Gorgier	1936	13.98
	Milvignes	8887	8.78
	Montalchez	244	6.42
	Peseux	5700	3.43
	Rochefort	1059	20.89
	Saint-Aubin-Sauges	2419	7.71
	Vaumarcus	262	1.78
	Total	38845	**105.62**

Mergers and name changes

The municipalities of Auvernier, Bôle and Colombier merged on 1 January 2013 into the new municipality of Milvignes.[3]

Demographics

Boudry District has a population (as of December 2011) of 38845.[1]

Most of the population (as of 2000) speaks French (31,695 or 86.3%) as their first language, German is the second most common (1,663 or 4.5%) and Italian is the third (1,007 or 2.7%). There are 21 people who speak Romansh.[4]

As of 2008, the population was 48.8% male and 51.2% female. The population was made up of 14,701 Swiss men (37.8% of the population) and 4,269 (11.0%) non-Swiss men. There were 16,480 Swiss women (42.4%) and 3,400 (8.8%) non-Swiss women.[5]

Of the population in the district, 7,421 or about 20.2% were born in Boudry and lived there in 2000. There were 12,122 or 33.0% who were born in the same canton, while 8,165 or 22.2% were born somewhere else in Switzerland, and 7,764 or 21.1% were born outside of Switzerland.[4]

As of 2000, there were 14,186 people who were single and never married in the district. There were 18,074 married individuals, 2,255 widows or widowers and 2,206 individuals who are divorced.[4]

There were 5,151 households that consist of only one person and 823 households with five or more people.

The historical population is given in the following chart:[6] [7]

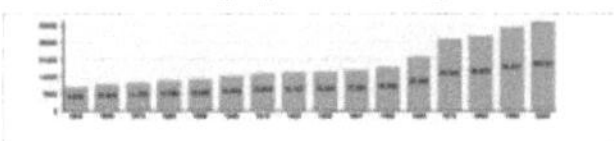

Politics

In the 2007 federal election the most popular party was the SP which received 24.74% of the vote. The next three most popular parties were the SVP (22.56%), the LPS Party (15.53%) and the FDP (14.97%). In the federal election, a total of 12,728 votes were cast, and the voter turnout was 51.9%.[8]

Religion

From the 2000 census, 11,032 or 30.0% were Roman Catholic, while 14,370 or 39.1% belonged to the Swiss Reformed Church. Of the rest of the population, there were 252 members of an Orthodox church (or about 0.69% of the population), there were 60 individuals (or about 0.16% of the population) who belonged to the Christian Catholic Church, and there were 2,289 individuals (or about 6.23% of the population) who belonged to another Christian church. There were 52 individuals (or about 0.14% of the population) who were Jewish, and 638 (or about 1.74% of the population) who were Islamic. There were 36 individuals who were Buddhist, 14 individuals who were Hindu and 41 individuals who belonged to another church. 7,447 (or about 20.28% of the population) belonged to no church, are agnostic or atheist, and 1,610 individuals (or about 4.38% of the population) did not answer the question.[4]

Education

In Boudry about 13,610 or (37.1%) of the population have completed non-mandatory upper secondary education, and 5,466 or (14.9%) have completed additional higher education (either university or a *Fachhochschule*). Of the 5,466 who completed tertiary schooling, 55.0% were Swiss men, 27.5% were Swiss women, 10.9% were non-Swiss men and 6.6% were non-Swiss women.[4]

In the canton of Neuchâtel most municipalities provide two years of non-mandatory kindergarten, followed by five years of mandatory primary education. The next four years of mandatory secondary education is provided at thirteen larger secondary schools, which many students travel out of their home municipality to attend.[9] During the 2010-11 school year, there were 40.5 kindergarten classes with a total of 758 students in District de Boudry. In the same year, there were 110 primary classes with a total of 2,135 students.[10]

References

[1] Swiss Federal Statistics Office – STAT-TAB (http://www.pxweb.bfs.admin.ch/Dialog/varval.asp?ma=px-d-01-2A01&ti=Ständige+und+Nichtständige+Wohnbevölkerung+nach+Region,+Geschlecht,+Nationalität+und++Alter&path=../Database/German_01 - Bevölkerung/01.2 - Bevölkerungsstand und -bewegung/&lang=1&prod=01&openChild=true&secprod=2) *Ständige und Nichtständige Wohnbevölkerung nach Region, Geschlecht, Nationalität und Alter* (German) accessed 3 October 2012

[2] Swiss Federal Statistical Office-Land Use Statistics (http://www.bfs.admin.ch/bfs/portal/de/index/themen/02/03/blank/data/gemeindedaten.html) 2009 data (German) accessed 25 March 2010

[3] Amtliches Gemeindeverzeichnis der Schweiz (http://www.bfs.admin.ch/bfs/portal/de/index/infothek/nomenklaturen/blank/blank/gem_liste/03.html) published by the Swiss Federal Statistical Office (German) accessed 2 January 2013

[4] STAT-TAB Datenwürfel für Thema 40.3 - 2000 (http://www.pxweb.bfs.admin.ch/Database/German_40 - EidgenÃ¶ssische VolkszÃ¤hlung/40.3 - 2000/40.3 - 2000.asp?lang=1&prod=40&secprod=3&openChild=true) (German) accessed 2 February 2011

[5] Canton of Neuchatel Statistics (http://www.ne.ch/neat/site/jsp/rubrique/rubrique.jsp?StyleType=bleu&DocId=40383), *République et canton de Neuchâtel - Recensement annuel de la population* (German) accessed 13 October 2011

[6] *Boudry* in German (http://www.hls-dhs-dss.ch/textes/d/D2823.php), French (http://www.hls-dhs-dss.ch/textes/f/F2823.php) and Italian (http://www.hls-dhs-dss.ch/textes/i/I2823.php) in the online *Historical Dictionary of Switzerland*.

[7] Swiss Federal Statistical Office STAT-TAB Bevölkerungsentwicklung nach Region, 1850-2000 (http://www.pxweb.bfs.admin.ch/Database/German_40 - EidgenÃ¶ssische VolkszÃ¤hlung/40.4 - 1850-2000/40.4 - 1850-2000.asp?lang=1&prod=40&secprod=4&openChild=true) (German) accessed 29 January 2011

[8] Swiss Federal Statistical Office, *Nationalratswahlen 2007: Stärke der Parteien und Wahlbeteiligung, nach Gemeinden/Bezirk/Canton* (http://www.bfs.admin.ch/bfs/portal/de/index/themen/17/02/blank/data/04/03.html) (German) accessed 28 May 2010

[9] EDK/CDIP/IDES (2010) Kantonale Schulstrukturen in der Schweiz und im Fürstentum Liechtenstein / Structures Scolaires Cantonales en Suisse et Dans la Principauté du Liechtenstein (http://edudoc.ch/record/35128/files/Schulsystem_alle.pdf) (Report). . Retrieved 24 June 2010.

[10] Statistical Department of the Canton of Neuchâtel (http://www.ne.ch/neat/site/jsp/rubrique/rubrique.jsp?CatId=9939) Mémento de l'année scolaire 2010/2011 (French) accessed 17 October 2011

Cantons_of_Switzerland

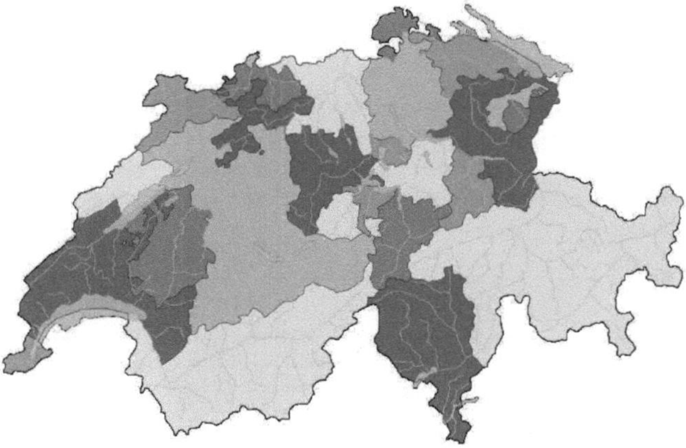

Valais

Ticino

Graubünden (Grisons)

Geneva

Vaud

Neuchâtel

Jura

Bern

Thurgau

Zurich

Aargau

Lucerne

Solothurn

Basel-Land

Schaffhausen

Uri

Schwyz

Glarus

St. Gallen

AI

AR

Obwalden

Nidwalden

Zug

Fribourg

Basel-Stadt

France

Italy

Liecht.

Austria

Germany

The 26 **cantons of Switzerland** are the member states of the federal state of Switzerland. Each canton was a fully sovereign state[1] with its own border controls, army and currency from the Treaty of Westphalia (1648) until the establishment of the Swiss federal state in 1848. The most recently created canton is the Canton of Jura, which separated from the Canton of Bern in 1979.[2]

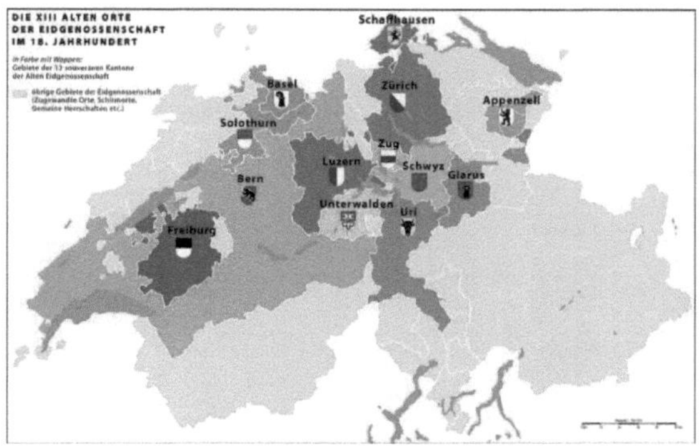

The 13 cantons of the Old Swiss Confederacy (1513-1798).

The name is derived from the French language word *canton* meaning *corner* or *district* (from which the term Cantonment is also derived).

History

In the 16th century, the Old Swiss Confederacy was composed of 13 sovereign cantons, and there were two different kinds: six land (or forest) cantons and seven city (or urban) cantons. Though they were technically part of the Holy Roman Empire, they had become *de facto* independent when the Swiss defeated Emperor Maximillian in 1499.[] The six forest cantons were democratic republics, whereas the seven urban cantons were oligarchic republics controlled by noble families.

Constitution

Each canton has its own constitution, legislature, government and courts.[3] Most of the cantons' legislatures are unicameral parliaments, their size varying between 58 and 200 seats. A few legislatures are general assemblies known as *Landsgemeinden*. The cantonal governments consist of either five or seven members, depending on the canton.[4] For the names of the institutions, see List of legislative and executive councils of the Cantons of Switzerland.

The Swiss Federal Constitution declares the cantons to be sovereign to the extent their sovereignty is not limited by federal law.[3] The cantons also retain all powers and competencies not delegated to the Confederation by the Constitution. Most significantly, the cantons are responsible for healthcare, welfare, law enforcement and public education; they also retain the power of taxation. The cantonal constitutions determine the degree of autonomy accorded to the municipalities, which varies but almost always includes the power to levy taxes and pass municipal laws. The sizes of the cantons vary from 37 km² to 7,105 km²; the populations vary from 15,471 to 1,244,400.

Direct democracy

As on the federal level, all cantons provide for (half-) direct democracy. Citizens may demand a popular vote to amend the cantonal constitution or laws, or to veto laws or spending bills passed by the parliament. General popular assemblies (*Landsgemeinde*) are now limited to the cantons of Appenzell Innerrhoden and Glarus. In all other cantons democratic rights are exercised by secret ballot.

List

The cantons are listed in the order given in the federal constitution.[a]

Coat of arms	Abbr	Canton	Since	Capital	Population[1]	Area (km²) [2]	Density (per km²) [3]	No. munic.[4]	Official languages
	ZH	**Zurich**	1351	Zurich	1406083		701	171	German
	BE	**Bern**	1353	Bern	985046		158	383	German, French
	LU	**Lucerne**	1332	Lucerne	381966		233	87	German
	UR	**Uri**	1291[5]	Altdorf	35382		33	20	German
	SZ	**Schwyz**	1291[5]	Schwyz	147904		143	30	German
	OW	**Obwalden**	1291[5] or 1315 (as part of Unterwalden)	Sarnen	35878		66	7	German
	NW	**Nidwalden**	1291[5] (as Unterwalden)	Stans	41311		138	11	German
	GL	**Glarus**	1352	Glarus	39217		51	3	German
	ZG	**Zug**	1352	Zug	116559		416	11	German
	FR	**Fribourg**	1481	Fribourg	284668		141	167	French, German
	SO	**Solothurn**	1481	Solothurn	259836		308	122	German
	BS	**Basel-Stadt**	1501 (as Basel until 1833/1999)	Basel	194090		5,072	3	German
	BL	**Basel-Landschaft**	1501/1833[6]	Liestal	277973		502	86	German
	SH	**Schaffhausen**	1501	Schaffhausen	77139		246	27	German
	AR	**Appenzell Ausserrhoden**	1513 [7]	Herisau[5]	53313		220	20	German
	AI	**Appenzell Innerrhoden**	1513[7]	Appenzell	15789		87	6	German
	SG	**St. Gallen**	1803[8]	St. Gallen	483101		222	85	German
	GR	**Graubünden**	1803[9]	Chur	193388		26	180	German, Romansh, Italian
	AG	**Aargau**	1803	Aarau	627893		388	220	German

	TG	**Thurgau**	1803[10]	Frauenfeld[6]	254528		229	80	German
	TI	**Ticino**	1803[11]	Bellinzona	336943		110	157	Italian
	VD	**Vaud**	1803[12]	Lausanne	725944		188	339	French
	VS	**Valais**	1815[13]	Sion	317022		53	143	French, German
	NE	**Neuchâtel**	1815/1857[14]	Neuchâtel	173183		206	53	French
	GE	**Geneva**	1815	Geneva	472530		1,442	45	French
	JU	**Jura**	1979[15]	Delémont	70542		82	64	French
	CH	**Switzerland**		Bern	7,968,705		174	2,596	German, French, Italian, Romansh

The two-letter abbreviations for Swiss cantons are widely used, e.g. on car license plates. They are also used in the ISO 3166-2 codes of Switzerland with the prefix "CH-" (*Confœderatio Helvetica*, Switzerland), e.g. CH-SZ for the canton of Schwyz.

Half-cantons

Six of the 26 cantons are traditionally, but no longer officially, called "half-cantons" (German: *Halbkanton*, French: *demi-canton*, Italian: *semicantone*), reflecting a history of mutual association or partition.

The half-cantons are identified in the first article of the Swiss Federal Constitution of 1999 by being joined to their other "half" with the conjunction "and":

> The People and the Cantons of Zurich, Bern, Lucerne, Uri, Schwyz, Obwalden and Nidwalden, Glarus, Zug, Fribourg, Solothurn, Basel-Stadt and Basel-Landschaft, Schaffhausen, Appenzell Ausserrhoden and Appenzell Innerrhoden, St. Gallen, Graubünden, Aargau, Thurgau, Ticino, Vaud, Valais, Neuchâtel, Geneva, and Jura form the Swiss Confederation.
>
> —Article 1 of the Federal Constitution of the Swiss Confederation (underlining not in original)[16]

The 1999 constitutional revision retained this distinction, on the request of the six cantonal governments, as a way to mark the historic association of the half-cantons to each other.[17] In contrast, the first article of the 1848 and 1874 constitutions constituted the Confederation as the union of "twenty-two sovereign cantons",[18] referring to the half-cantons as "Unterwalden (above and beneath the woods)", "Basel (city and country)" and "Appenzell (both Rhoden)".[19] While the older constitutions referred to these states as "half-cantons", a term that remains in popular use, the 1999 revision and official terminology since then use the appellation "cantons with half a cantonal vote".[20]

With their mutual association a purely historical matter, the half-cantons are since 1848 equal to the other cantons in all but two respects:[21]

- They elect only one member of the Council of States instead of two (Cst. art. 150 par. 2).
- In popular referendums about constitutional amendments, which require for adoption a national popular majority as well as the assent of a majority of the cantons (*Ständemehr / majorité des cantons*), the result of the half-cantons' popular vote counts only one half of that of the other cantons (Cst. arts. 140, 142). This means that for purposes of a constitutional referendum, at least 12 out of a total of 23 cantonal popular votes must support the amendment.[22]

Caricature of the division of Basel, 1833

The reasons for the association between the three pairs of half-cantons are varied:

- *Unterwalden* never consisted of a single unified jurisdiction. Originally, Obwalden, Nidwalden, and the Abbey of Engelberg formed distinct communities. The collective term *Unterwalden* remains in use, however, for the area that partook in the creation of the original Swiss confederation in 1291 with Uri and Schwyz. The Federal Charter of 1291 called for representatives from each of the three "areas".[23] [24]
- The canton of Appenzell divided itself into an "inner" and "outer" half ("Rhoden") as a consequence of the Reformation in Switzerland in 1597:[25] Appenzell Innerrhoden (Catholic) and Appenzell Ausserrhoden (Protestant).
- The canton of Basel divided itself as a consequence of a revolt of the Basel countryside in 1833, in order to promote equality among its citizenry, combating claims between rural and city residents over preferential status:[26] Basel-Landschaft and Basel-Stadt.

Names in national languages

(Names appear in bold when corresponding to the cantonal official language)

Abbr	Common English	Other English forms	German	French	Italian	Romansh
AG	Aargau	Argovia	**Aargau**	Argovie	Argovia	Argovia
AI	Appenzell Innerrhoden	Appenzell Inner-Rhodes	**Appenzell Innerrhoden**	Appenzell Rhodes-Intérieures	Appenzello Interno	Appenzell dadens
AR	Appenzell Ausserrhoden	Appenzell Outer-Rhodes	**Appenzell Ausserrhoden**	Appenzell Rhodes-Extérieures	Appenzello Esterno	Appenzell dador
BS	Basel-Stadt	Basle-City	**Basel-Stadt**	Bâle-Ville	Basilea-Città	Basilea-Citad
BL	Basel-Landschaft	Basle-Country	**Basel-Landschaft**	Bâle-Campagne	Basilea-Campagna	Basilea-Champagna
BE	Bern	Berne	**Bern**	**Berne**	Berna	Berna
FR	Fribourg	Friburg	**Freiburg**	**Fribourg**	Friborgo	Friburg
GE	Geneva	-	Genf	**Genève**	Ginevra	Genevra
GL	Glarus	Glaris	**Glarus**	Glaris	Glarona	Glaruna
GR	Graubünden	Grisons	**Graubünden**	Grisons	**Grigioni**	**Grischun**
JU	Jura	-	Jura	**Jura**	Giura	Giura
LU	Lucerne	-	**Luzern**	Lucerne	Lucerna	Lucerna
NE	Neuchâtel	-	Neuenburg	**Neuchâtel**	Neuchâtel	Neuchâtel
NW	Nidwalden	Nidwald	**Nidwalden**	Nidwald	Nidvaldo	Sutsilvania
OW	Obwalden	Obwald	**Obwalden**	Obwald	Obvaldo	Sursilvania
SH	Schaffhausen	Schaffhouse	**Schaffhausen**	Schaffhouse	Sciaffusa	Schaffusa
SZ	Schwyz	-	**Schwyz**	Schwyz (or Schwytz)	Svitto	Sviz
SO	Solothurn	Soleure	**Solothurn**	Soleure	Soletta	Soloturn
SG	St. Gallen	St. Gall	**St. Gallen**	Saint-Gall	San Gallo	Son Gagl

TG	Thurgau	Thurgovia	**Thurgau**	Thurgovie	Turgovia	Turgovia
TI	Ticino	Tessin	Tessin	Tessin	**Ticino**	Tessin
UR	Uri	-	**Uri**	Uri	Uri	Uri
VS	Valais	Wallis	**Wallis**	**Valais**	Vallese	Vallais
VD	Vaud	-	Waadt	**Vaud**	Vaud	Vad
ZG	Zug	-	**Zug**	Zoug	Zugo	Zug
ZH	Zurich	-	**Zürich**	Zurich	Zurigo	Turitg

Admission of new cantons

The enlargement of Switzerland by way of the admission of new cantons ended in 1815. After a failed attempt of Vorarlberg to join Switzerland in 1919, the idea of resuming Swiss enlargement was revived in 2010 by a parliamentary motion that would allow the accession of regions bordering on Switzerland.

See also

- List of legislative and executive councils of the Cantons of Switzerland
- Data codes for Switzerland#Cantons
- List of cantons of Switzerland by elevation
- Flags of Swiss cantons
- Cantonal bank, a commercial bank (at least partially) owned by the canton

Notes

1. This is the order generally used in Swiss official documents. At the head of the list are the three city cantons that were considered preeminent in the Old Swiss Confederacy; the other cantons are listed in order of accession to the Confederation. This traditional order of precedence among the cantons has no practical relevance in the modern federal state, in which the cantons are equal to one another, although it still determines formal precedence among the cantons' officials (see Swiss order of precedence).
2. as of 5 April 2009
3. km²
4. Per km², based on 2000 population
5. As of 31 December 2007,
6. Seat of government and parliament is Herisau, the seat of the judicial authorities is Trogen
7. Seat of parliament half-yearly alternates between Frauenfeld and Weinfelden

References

- (German). Cited as *Ehrenzeller*.
- Cited as *Häfelin*.

[1] *Cantons, In the Old Confederation until 1798* in German (http://www.hls-dhs-dss.ch/textes/d/D26414-1-1.php), French (http://www.hls-dhs-dss.ch/textes/f/F26414-1-1.php) and Italian (http://www.hls-dhs-dss.ch/textes/i/I26414-1-1.php) in the online *Historical Dictionary of Switzerland.*

[2] *Jura (Canton)* in German (http://www.hls-dhs-dss.ch/textes/d/D7399.php), French (http://www.hls-dhs-dss.ch/textes/f/F7399.php) and Italian (http://www.hls-dhs-dss.ch/textes/i/I7399.php) in the online *Historical Dictionary of Switzerland.*

[3] *Cantons, In the Federal State since 1848* in German (http://www.hls-dhs-dss.ch/textes/d/D26414-1-4.php), French (http://www.hls-dhs-dss.ch/textes/f/F26414-1-4.php) and Italian (http://www.hls-dhs-dss.ch/textes/i/I26414-1-4.php) in the online *Historical Dictionary of Switzerland.*

[4] Swiss Government website (http://www.ch.ch/karte/index.html?lang=en) with links to each cantonal government, accessed 11 November 2008
[5] founding forest-canton, foundation date traditionally given as either 1307, 1304 or 1291 (see Foundation of the Old Swiss Confederacy).
[6] part of Basel until 1833/1999
[7] part of Appenzell until 1597/1999
[8] Act of Mediation, formed out of the Canton of Säntis and the northern half of the Canton of Linth.
[9] Act of Mediation; formerly the Canton of Raetia, comprising the earlier Three Leagues.
[10] coterminous with the canton of Thurgau of the Helvetic Republic (1798), formerly a condominium.
[11] combining the former cantons of Bellinzona and Lugano; see *Ennetbirgische Vogteien*.
[12] Act of Mediation, formerly Canton of Léman.
[13] Restoration, formerly the Simplon *département*
[14] claimed by Frederick William III of Prussia until the Neuchâtel Crisis of 1856–1857.
[15] seceded from Berne
[16] Federal Constitution of the Swiss Confederation of 18 April 1999, SR/RS 101 (E (http://www.admin.ch/ch/e/rs/c101.html), D (http://www.admin.ch/ch/d/sr/c101.html), F (http://www.admin.ch/ch/f/rs/c101.html), I (http://www.admin.ch/ch/i/rs/c101.html)), art. 1 (E (http://www.admin.ch/ch/e/rs/101/a1.html), D (http://www.admin.ch/ch/d/sr/101/a1.html), F (http://www.admin.ch/ch/f/rs/101/a1.html), I (http://www.admin.ch/ch/i/rs/101/a1.html))
[17] Felix Hafner / Rainer J. Schweizer in *Ehrenzeller*, Art. 1 N 2; Häfelin, N 966.
[18] Twenty-three after the creation of the Canton of Jura in 1978.
[19] Bundesverfassung der Schweizerischen Eidgenossenschaft vom 29. Mai 1874 (http://www.verfassungen.de/ch/verf74-i.htm), Bundesverfassung der Schweizerischen Eidgenossenschaft vom 12. September 1848 (http://www.verfassungen.de/ch/verf48.htm) (German); author's translation.
[20] Felix Hafner / Rainer J. Schweizer in *Ehrenzeller*, Art. 1 N 10; Häfelin, N 963
[21] Häfelin, N 963, 967
[22] Häfelin, N 950
[23] Pacte fédéral du 1er (http://www.admin.ch/org/polit/00056/index.html?lang=fr) août 1291] sur Admin.ch "vallée inférieure d'Unterwald" signifie Nidwald.
[24] Pacte fédéral du 1er août 1291 (http://hypo.ge.ch/www/cliotexte//html/suisse.histoire.1291.html) sur Cliotexte
[25] Réforme catholique, Contre-Réforme et scission (http://hls-dhs-dss.ch/textes/f/F7389-1-10.php) Article du dictionnaire historique de la Suisse
[26] De la République helvétique à la division du canton (1798-1833) (http://hls-dhs-dss.ch/textes/f/F7387-2-2.php) Article du dictionnaire historique de la Suisse

External links

- Swissworld.org (http://www.swissworld.org/en/politics/general_information/the_cantons/) – The cantons of Switzerland
- GeoPuzzle (http://www.geomit.ch/GeoPuzzles/SchweizPuzzle/schweiz.html) – Assemble cantons on a Swiss map
- Badac (http://www.badac.ch) – Database on Swiss cantons and cities (French/German)

Canton_of_Neuchâtel

County (Principality) of Neuchâtel
Grafschaft (Fürstentum) Neuenburg **(de)**
Comté (Principauté) de Neuchâtel **(fr)**

State of the Holy Roman Empire (to 1648)
Associate of the Old Swiss Confed. (from 1406)

← 1034–1848 →

Coat of arms

Capital		Neuchâtel
Government		Principality
Historical era		Middle Ages
-	City founded	1011
-	County founded	1034
-	Became associate of Old Swiss Confed.	from 1406
-	Inherited by Orléans-Longueville	1504
-	Elected to Prussia	1707
-	French occupation	1806–14
-	Joined Swiss Confed. as canton	1815
-	Neuchâteloise revolution	1 March 1848

Neuchâtel (French: *Canton de Neuchâtel*, IPA: [kɑ̃tɔ̃ də nøʃɑtɛl]) is a canton of French speaking western Switzerland. In 2007, its population was 169,782 of which 39,654 (or 23.4%) were foreigners.[1] The capital is Neuchâtel.

History

The only part of present-day Switzerland to enter the Confederation as a principality (in 1814), Neuchâtel has a unique history. Its first recorded ruler, Rudolph III of Burgundy mentioned Neuchâtel in his will in 1032. The dynasty of Count Ulrich von Fenis took over the town and its territories in 1034. The dynasty prospered and by 1373 all the lands now part of the canton belonged to the count. In 1405, the cities of Bern and Neuchâtel entered a union. The lands of Neuchâtel passed to the lords of Freiburg im Breisgau (Germany) about a century later, and then in 1504 to the French house of Orléans-Longueville (Valois-Dunois). Neuchâtel's Swiss allies then occupied it from 1512-1529 before returning it to its widowed Countess Jeanne de Hochberg.

The French preacher Guillaume Farel brought the teachings of the Protestant Reformation to the area in 1530. When the house of Orléans-Longueville became extinct with Marie d'Orléans-Longueville's death in 1707, the Principality of Neuchâtel (German: *Fürstentum Neuenburg*) went to King Frederick I in Prussia of the Berlin-based Hohenzollern dynasty, who then ruled Neuchâtel in personal union. Napoléon Bonaparte deposed King Frederick William III of Prussia as prince of Neuchâtel and appointed instead his chief of staff Louis Alexandre Berthier. In the course of the Napoleonic Wars the principality provided for a rangers battalion within Napoléon's Grande Armée since 1807. The rangers were nicknamed *Canaris* (i.e. Canaries) because of their yellow uniforms.

Le Locle, 1907

After the Liberation Wars the principality was restored to Frederick William III in 1814.[2] The *Conseil d'Etat* (state council, i.e. government of Neuchâtel) addressed him in May 1814 requesting the permission to establish a special battalion, a *Bataillon de Chasseurs*, for the service of his majesty.[2] Frederick William III then established by his most-supreme cabinet order (allerhöchste Cabinets-Ordre), issued in Paris on 19 May 1814, the Bataillon des Tirailleurs de la Garde following the same principles as with the Neuchâtel battalion within the Grande Armée.[2] The Conseil d'Etat of Neuchâtel had the right of nomination for the battalion's officers. The commander was the battalion's only officer chosen by the monarch.

A year later he agreed to allow the principality to join the Swiss Confederation, then not yet an integrated federation, but a confederacy, as a full member. Thus Neuchâtel became the first and only monarchy to join the otherwise entirely republican Swiss cantons. This situation changed in 1848 when a peaceful revolution took place and established a republic, in the same year that the modern Swiss Confederation was transformed into a federation. King Frederick William IV of Prussia did not give in immediately and several attempts at counter-revolution took place, culminating in the Neuchâtel Crisis of 1856–57. In 1857, Frederick William renounced his claims on the area.

Geography

View of Lake Neuchâtel from the northern shore, port of Vaumarcus

The canton of Neuchâtel is located in Romandy, the western part of Switzerland, it is also located in the Jura mountainous region. To its northeast it borders the canton of Bern, to the northwest France. Lake Neuchâtel lies southeast of the canton, while the canton of Vaud is southwest of the canton of Neuchâtel. The canton lies in the central area of the Jura Mountains. Lake Neuchâtel drains the lands in the south, whilst the River Doubs drains the northern areas.

The canton is commonly divided into three regions. The viticultural region is located along the lake. Its name derives from the many vineyards found there. The region called *Les Vallées* lies further north. The two largest valleys of the canton of Neuchâtel lie in this region: the *Ruz Valley* and the *Val de Travers*. Both valleys lie at about 700 m (ft). The highest region of the canton, however, is the *Neuchâtelois Mountains* at 900 m (ft) to 1065 m (ft). This region is made up of a long valley home to La Chaux-de-Fonds, Le Locle and La Brévine.

Government

Château de Neuchâtel, now seat of the cantonal government

Neuchâtel was one of the first cantons in Switzerland to grant women the vote (1959) and also to grant the vote to foreigners holding a residence permit and who have been domiciled in the canton for at least 5 years (2002), as well as to lower the voting age to 18.

The legislature, the Grand Council of Neuchâtel, has 115 seats distributed in proportion to the population of the 6 districts that make up the electoral constituencies: Neuchâtel (35 seats), Boudry, (25) Val-de-Travers (8), Val-de-Ruz (10), Le Locle (10), La Chaux-de-Fonds (27). The State Council (cantonal government), 5 "ministers" who assume the annual presidency in turn and manage the departments of justice, health and safety; finance and social welfare; public economy; regional management; education and culture. The cantonal authorities, which have their seat in the castle (the Château de Neuchâtel), are elected every four years by universal suffrage.

The people also elect their representatives to the federal parliament every four years: five of the 200 members of the National Council (lower chamber) and 2 of the 46 members of the Council of States (upper chamber).

Political subdivisions

Districts

The Canton is divided into 6 districts:

- Boudry with capital Boudry
- La Chaux-de-Fonds with capital La Chaux-de-Fonds
- Le Locle with capital Le Locle
- Neuchâtel with capital Neuchâtel
- Val-de-Ruz with capital Cernier
- Val-de-Travers with capital Val-de-Travers

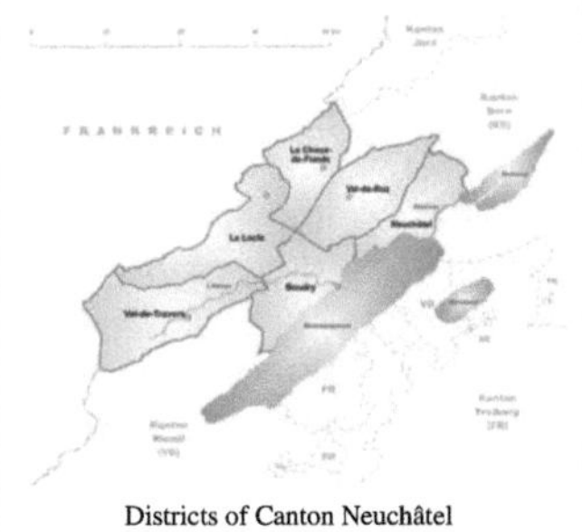

Districts of Canton Neuchâtel

Municipalities

There are 53 municipalities in the canton (As of 2009).[]

Demographics

La Chaux-de-Fonds, most populous city in the canton

The population is almost entirely French-speaking. The canton has historically been strongly Protestant, but in recent decades it has received an influx of Roman Catholic arrivals. In 2000, its population was closely split between Protestants (38%) and Roman Catholics (31%).[3]

The 173183 inhabitants (as of 2011)[4] are fairly evenly distributed with many small towns and villages lining the shore of the Lake of Neuchâtel The average population density is 209 people per km^2 (542 sq mi). Neuchâtel (2011 population: 33412) is the canton's capital while La Chaux-de-Fonds (2011 population: 37843) is the canton's largest settlement. Some 38,000 of the inhabitants, or a little less than a quarter of the population, are of foreign extraction.

Economy

The canton is well known for its wines, which are grown along the Lake Neuchâtel shore, and for its absinthe. The Val-de-Travers is famous as the birthplace of absinthe, which has now been re-legalized both in Switzerland and globally. There are dairy farming and cattle breeding in the valleys, but it is for the breeding of horses that Neuchâtel has a fine reputation. Watchmaking, see "Watch Valley" [5], is well established in the canton, with fine mechanics and microchip production being established more recently. Higher educational institutions include Haute école de gestion de Neuchâtel and the University of Neuchâtel.

Notes and references

[2] Ilse Nicolas, „Militaria: Die Neffschandeller am Schlesischen Busch“, in: Ilse Nicolas, *Kreuzberger Impressionen* ([1]1969), Berlin: Haude & Spener, [2]1979, (=Berlinische Reminiszenzen; vol. 26), pp. 111–114, here p. 111. ISBN 3-7759-0205-8.

[4] Swiss Federal Statistics Office – STAT-TAB (http://www.pxweb.bfs.admin.ch/Dialog/varval.asp?ma=px-d-01-2A01&ti=Ständige+und+Nichtständige+Wohnbevölkerung+nach+Region,+Geschlecht,+Nationalität+und++Alter&path=../Database/German_01 - Bevölkerung/01.2 - Bevölkerungsstand und -bewegung/&lang=1&prod=01&openChild=true&secprod=2) *Ständige und Nichtständige Wohnbevölkerung nach Region, Geschlecht, Nationalität und Alter* (German) accessed 3 October 2012

[5] http://www.watchvalley.ch/e/home/

External links

- Official website (http://www.ne.ch/)
- Official statistics (http://www.statistik.admin.ch/stat_ch/ber00/ekan_ne.htm)
- Neuchâtel and surroundings "Watch Valley" (http://www.juraregion.ch/en/discoveries/watch-valley.html:)
- Portal Neuch.ch (http://www.neuch.ch)
- Village of Valangin (http://www.switzerland-traveler.com/Valangin/Valangin.html)
- International Watchmaking Museum (http://www.bargaintraveleurope.com/12/Switzerland_Watchmaking_Museum_La_Chaux-de-Fonds.htm)

Coat_of_arms

Escutcheon

Field

Supporter

Supporter

Motto (Scotland)

Crest

Torse

Mantling

Helm

Coronet

Compartment

Order

Ordinaries

Common

charges

Motto

Achievement elements

A **coat of arms** is a unique heraldic design on an escutcheon (i.e. shield), surcoat, or tabard. A surcoat, and subsequently a *coat of arms* was used by medieval knights to cover, protect, and identify the wearer. Thus these are sometimes termed *coat armory*. The coat of arms on an escutcheon forms the central element of the full heraldic achievement which consists of shield, supporters, crest, and motto. The design is a symbol unique to an individual person or family, corporation, or state. Such displays are commonly called *armorial bearings*, *armorial devices*, *heraldic devices*, or simply *arms*.

The ancient Romans used similar insignias similar on their shields, but these identified military units rather than individuals. The first evidence of medieval coats of arms is found in the Bayeux Tapestry from the 11th Century, where some of the combatants carry shields painted with crosses. Coats of arms came into general use by feudal lords and knights in battle in the 12th Century. By the 13th Century arms had spread beyond their initial battlefield use to become a kind of flag or logo for families in the higher social classes of Europe, inherited from one generation to the next. Exactly who had a right to use arms, by law or social convention, varied to some degree between countries. In the German-speaking region both the aristocracy and *burghers* (non-noble free citizens) used arms, while in most of the rest of Europe they were limited to the aristocracy. The use of arms spread to Church clergy, and to towns as civic identifiers, and to royally-chartered organizations such as universities and trading companies. Flags developed from coats of arms, and the arts of vexillology and heraldry are closely related. The coats of arms granted to commercial companies are a major source of the modern logo.

Despite no widespread regulation, and even with a lack in many cases of national regulation, heraldry has remained rather consistent across Europe, where traditions alone have governed the design and use of arms. Unlike seals and other general emblems, heraldic achievements have a formal description called a blazon, expressed in a jargon that allows for consistency in heraldic depictions.

In the 21st century, coats of arms are still in use by a variety of institutions and individuals; for example, many European cities and universities have guidelines on how their coats of arms may be used, and protect their use as trademarks.[1] [2] [3] Many societies exist that also aid in the design and registration of personal arms. Some nations, like England and Scotland, still maintain the same heraldic authorities which have traditionally granted and regulated arms for centuries and continue to do so in the present day.

Traditions and usage

In the heraldic traditions of England and Scotland an individual, rather than a family, had a coat of arms. In those traditions coats of arms are legal property transmitted from father to son; wives and daughters could also bear arms modified to indicate their relation to the current holder of the arms. Undifferenced arms are used only by one person at any given time. Other descendants of the original bearer could bear the ancestral arms only with some difference: usually a color change or the addition of a distinguishing charge. One such charge is the label, which in British usage (outside the Royal Family) is now always the mark of an heir apparent or (in Scotland) an heir presumptive.

The German *Hyghalmen Roll*, ca. late 15th century, illustrates the German practice of thematic repetition from the arms in the crest

Because of their importance in identification, particularly in seals on legal documents, the use of arms was strictly regulated; few countries continue in this today. This has been carried out by heralds and the study of coats of arms is therefore called "heraldry". Some other traditions (e.g., Polish heraldry) are less restrictive — allowing, for example, all members of a dynastic house or family to use the same arms, although one or more elements may be reserved to the head of the house.

In time, the use of arms spread from military entities to educational institutes, and other establishments.[2] According to a design institute article, "The modern logo and corporate livery have evolved from the battle standard and military uniform of medieval times".[4]

In his book, *The Visual Culture of Violence in the Late Middle Ages*, Valentin Groebner argues that the images composed on coats of arms are in many cases designed to convey a feeling of power and strength, often in military terms.[5] The author Helen Stuart argues that some coats of arms were a form of corporate logo.[6] Museums on medieval armory also point out that as emblems they may be viewed as precursors to the corporate logos of modern society, used for group identity formation.[7] [8]

When knights were so encased in armour that no means of identifying them was left, the practice was introduced of painting their insignia of honour on their shield as an easy method of distinguishing them. Originally these were granted only to individuals, but were afterward made hereditary in England by King Richard I, during his crusade to Palestine.

European tradition

British heraldry

In Scotland, the Lord Lyon King of Arms has criminal jurisdiction to enforce the laws of arms. In England, Northern Ireland and Wales the use of arms is a matter of civil law and regulated by the College of Arms and the Court of Chivalry.

The coat of arms of the University of Plymouth, which were granted in 2008 by the College of Arms, shows the heater shield and style popular in modern British heraldry.[9]

In reference to a dispute over the exercise of authority over the Officers of Arms in England, Arthur Annesley, 1st Earl of Anglesey, Lord Privy Seal, declared on 16 June 1673 that the powers of the Earl Marshal were "to order, judge, and determine all matters touching arms, ensigns of nobility, honour, and chivalry; to make laws, ordinances, and statutes for the good government of the Officers of Arms; to nominate Officers to fill vacancies in the College of Arms; to punish and correct Officers of Arms for misbehaviour in the execution of their places". It was further declared that no patents of arms or any ensigns of nobility should be granted and no augmentation, alteration, or addition should be made to arms without the consent of the Earl Marshal.

Irish heraldry

In Ireland, since 1552, the usage and granting of coats of arms has been strictly regulated by the Government of Ireland, through the Office of the Chief Herald of Ireland.

German and Scandinavian heraldry

The heraldic tradition and style of modern and historic Germany and the Holy Roman Empire — including national and civic arms, noble and burgher arms, ecclesiastical heraldry, heraldic displays and heraldic descriptions — stand in contrast to Gallo-British, Latin and Eastern heraldry, and strongly influenced the styles and customs of heraldry in the Nordic countries, which developed comparatively late.[10]

Coat of arms of the city of Vaasa, showing the shield with the Royal House of Wasa emblem, a crown and a Cross of Liberty pendant.

In the Nordic countries, provinces, regions, cities and municipalities have a coat of arms. These are posted at the borders and on buildings containing official offices, as well as used in official documents and on the uniforms of municipal officers. Arms may also be used on souvenirs or other effects, given that an application has been granted by the municipal council.

Other European countries

At a national level, "coats of arms" were generally retained by European states with constitutional continuity of more than a few centuries, including constitutional monarchies like Denmark as well as old republics like San Marino and Switzerland.

In Italy the use of coats of arms was only loosely regulated by the states existing before the unification of 1860. Since the Consulta Araldica, the college of arms of the Kingdom of Italy, was abolished in 1948, personal coats of arms and titles of nobility, though not outlawed, are not recognised.

Among the states ruled by communist regimes, emblems resembling the Soviet design were adopted in all the Warsaw Pact states except Czechoslovakia and Poland.

Since 1989, some of the ex-Communist states, as Romania, have resumed their former arms, often with only the symbols of monarchy removed.

Asia and Africa

Arab world

Flag of the Ottoman Empire (1844–1924)

Egyptian coats of arms showing common Near and Middle Eastern motifs, namely the crescent and stars which are symbols of the region's predominant religion, Islam, and an eagle.

With the formation of the modern nation states of the Arab World in the second half of the 20th century, European traditions of heraldry were partially adopted for state emblems.

These emblems often involve the star and crescent symbol taken from the Ottoman flag. Another commonly seen charge is the eagle, which is a symbol attributed to Saladin, and the hawk of the Qureish.

Mitsubishi logo

Japanese Mon

Japanese emblems, called *kamon* (often abbreviated "mon"), are family badges which often date back to the 7th century, and are used in Japan today. The Japanese tradition is independent of the European, and thus very different in style; but as in Europe many abstract and floral elements are used.

Yet, even these simple designs often express an origin. An example in recent use is the logo of Mitsubishi corporation which started as a shipping and maritime enterprise and whose emblem is based on a water chestnut derived from its maritime history with a military naval influence. The word *mitsu* means the number 3 and the word *hishi* meaning "water chestnut" (pronounced *bishi* in some combinations; see rendaku) originated from the emblem of the warrior Tosa Clan. The battleships of the Tosa Clan had been used in the late 19th century in the First Sino-Japanese War to reach Korea[11] and their name was given to a modern battleship. The Tosa water chestnut leaf mon was then drawn as a rhombus or diamond shape in the Mitsubishi logo.[12] [13]

New World practices

Canada

The Queen of Canada has delegated her prerogative to grant armorial bearings to the Governor General of Canada. Canada has its own Chief Herald and Herald Chancellor. The Canadian Heraldic Authority is situated at Rideau Hall.[] []

United States

The Great Seal of the United States uses on the obverse as its central motif an heraldic achievement described as being the arms of the nation.[14] The seal, and the armorial bearings, were adopted by the Continental Congress on 20 June 1782, and is a shield divided palewise into thirteen pieces, with a blue chief, which is displayed upon the breast of an American bald eagle. The crest is thirteen stars breaking through a glory and clouds, displayed with no helm, torse or mantling (unlike most European precedents).

Only a few of the American states have adopted a coat of arms, which is usually designed as part of the respective state's seal. Vermont has both a state seal and a state coat of arms that are independent of one another (though both contain a pine tree, a cow and sheaves of grain); the seal is used to authenticate documents, whilst the heraldic device represents the state itself.

Ecclesiastic practice

Catholic Church

Vatican City State and the Holy See each have their own coat of arms. As the Papacy is not hereditary, its occupants display their personal arms combined with those of their office.

Some Popes came from armigerous (noble) families; others adopted coats of arms during their career in the Church. The latter typically allude to their ideal of life, or to specific Pontifical programmes.[15] A well-known and widely displayed example in recent times was Pope John Paul II's arms. His selection of a large letter M (for the Virgin Mary) was intended to express the message of his strong Marian devotion.[16]

Coat of arms of Pope John Paul II, display the Papal tiara and crossed keys of the pontifical office.

Roman Catholic dioceses also are assigned a coat of arms, as do a basilica or papal church also gets a coat of arms, the latter usually displaying these on the building. These may be used in countries which otherwise do not use heraldic devices.

In countries like Scotland with a strong statutory heraldic authority, arms will need to be officially granted and recorded.

Flags and banners

Flags are used to identify ships (where they are called ensigns), embassies and such, and they use the same colors and designs found in heraldry, but they are not usually considered to be heraldic. A country may have both a national flag and a national coat of arms, and the two may not look alike at all. For example, the flag of Scotland (St Andrew's Cross) has a white saltire on a blue field, but the royal arms of Scotland has a red lion within a double tressure on a gold (or) field.

See also

- National emblem
- Seal
- Insignia
- List of coats of arms
- Baron and feme
- Siebmachers Wappenbuch (Coats of arms from Germany, Switzerland, Austria, etc.)

Notes

[1] Glasgow University - The Coat of Arms (http://www.universitystory.gla.ac.uk/coat-of-arms/)
[2] Educational Institute Coat of arms (http://www.emeraldinsight.com/Insight/ViewContentServlet?Filename=Published/EmeraldFullTextArticle/Articles/1680100103.html)
[3] Policy on use of the Workmark and Insignia of McGill University (http://www.mcgill.ca/files/secretariat/WordmarkandInsigniaofMcGillUniversity-Policyontheuseof.pdf)
[4] Understanding Corporate Identity (http://www.dia.org.au/content.cfm?news=30&id=102)
[5] Valentin Groebner, 2004, *The Visual Culture of Violence in the Late Middle Ages* ISBN 978-1-890951-37-5
[6] *Employee Identification with the Corporate Identity* International Studies of Management and Organization, Volume 32, Number 3, 2002
[7] Group Identity Formation in the German Renaissance (http://www.phil-hum-ren.uni-muenchen.de/GermLat/Acta/Bernstein.htm)
[8] European Coat of Arms Museum (http://www.wien.gv.at/english/culture/museums/coat-of-arms.htm)
[9] College of Arms (http://www.college-of-arms.gov.uk/Newsletter/018.htm)
[10] ISBN 0-7137-0940-5 p.129.
[11] Donald Calman, 1992 *The Nature and Origins of Japanese Imperialism* ISBN 0-415-06710-3
[12] Mitsubishi Mark (http://www.mitsubishi.com/e/group/mark.html)
[13] Mitsubishi History (http://www.mitsubishi.com/e/history/index.html)
[14] 2004 Seal Broch (http://www.state.gov/documents/organization/27807.pdf)
[15] Coat of Arms of His Holiness Benedict XVI (http://www.vatican.va/holy_father/benedict_xvi/elezione/stemma-benedict-xvi_en.html)
[16] Vatican press office (http://www.vatican.va/news_services/press/documentazione/documents/sp_ss_scv/insigne/sp_ss_scv_stemma-bandiera-sigillo_en.html)

References

- This article incorporates text from a publication now in the public domain: "Pimbley's dictionary of heraldry", Arthur Francis Pimbley (1908)

External links

- College of Arms (http://www.college-of-arms.gov.uk/): Repository of the coats of arms and pedigrees of English, Welsh, Northern Irish and Commonwealth families and their descendants together with, and in principle under the control of, the legal body the Court of Chivalry, both medieval in origin.
- The Court of the Lord Lyon (http://www.lyon-court.com/lordlyon/CCC_FirstPage.jsp): the statutory heraldry office for Scotland
- The Armorial Register, International Register of Arms (http://www.armorial-register.com): A fully illustrated record of contemporary Coats of Arms
- Free access (http://www.luz-herald.net/free/) to Burke's General Armory (incomplete, 1500 British surnames), Pimbley's Dictionary of Heraldry and Blason des familles d'Europe, Grand Armorial Universel (15,000 European surnames)
- Royal Dutch Library page for the "Wapenboek Beyeren" (http://www.kb.nl/galerie/wapenboek/index.html/) written by Claes Heynenzoon around 1400, containing over 1000 drawings of coats of arms
- General armorial of noble families in the Russian Empire (Gerbovnik) (http://gerbovnik.com)

Blazon

In heraldry and heraldic vexillology, a **blazon** is a formal description of a coat of arms, flag or similar emblem, from which the reader can reconstruct the appropriate image. The visual depiction of a coat of arms or flag traditionally has considerable latitude in design, while a blazon specifies the essentially distinctive elements; thus it can be said that a coat of arms or flag is primarily defined not by a picture but rather by the wording of its blazon (though often flags are in modern usage additionally and more precisely defined using geometrical specifications). *Blazon* also refers to the specialized language in which a blazon is written, and, as a verb, to the act of writing such a description. This language has its own vocabulary, grammar and syntax, or rules governing word order, which becomes essential for comprehension when blazoning a complex coat of arms.

Other objects, such as badges, banners, and seals may be described in blazon.

Etymology

The word blazon is derived from French *blason*, "shield." It is found in English by the end of the 14th century.[1] Formerly, experts in heraldry assumed that the word was related to the German verb *blasen*, "to blow (a horn)."[2] Present-day lexicographers reject this theory as conjectural and disproved.[1]

Grammar

The blazon of armorials follows a rigid formula:

- Every blazon of a coat of arms begins by describing the field (background), followed by a comma ",". In a majority of cases this is a single tincture; e.g. *Azure* (blue). If the field is complex, the variation is described, followed by the tinctures used; e.g. *Chequy gules and argent* (checkered red and white). If the shield is divided, the division is described, followed by the tinctures of the subfields, beginning with the dexter side (shield bearer's right, but viewer's left) of the *chief* (upper) edge; e.g. *Party per pale argent and vert* (dexter half silver, sinister half green), or *Quarterly argent and gules* (clockwise from top left: white, red, white, red).
- The principal charge(s) are then named, with their tincture(s); e.g. *a bend Or.*
- The principal charge is followed by any other charges placed around or on it. If a charge be a bird or beast, its attitude is described, followed by the animal's tincture, followed by anything that may be differently coloured; e.g. *An eagle **displayed** gules, **armed** and wings charged with trefoils Or* (see the coat of arms of Brandenburg).

Any accessories present — such as crown/coronet, helmet, torse, mantling, crest, motto, supporters and compartment — are then described in turn, using the same terminology and syntax.

Azure, a bend Or — a coat made famous by the court case *Scrope v. Grosvenor*

Party per pale argent and vert, a tree eradicated counterchanged (arms of Behnsdorf)

Quarterly argent and gules

Argent, an eagle displayed gules armed and wings charged with trefoils Or (arms of Brandenburg)

A composite shield is blazoned one panel at a time, proceeding by rows from chief (top) to base, and within each row from dexter (the right side of the bearer holding the shield) to sinister; in other words, from the viewer's left to right. A divided shield is blazoned "party per [line of division]" or "**parted** per [line of division]", though the word "party" or "parted" is almost always omitted (e.g. *"Per pale argent and vert, a tree eradicated counterchanged"*). A tincture is sometimes replaced by "of the first", "of the second" etc. to avoid repetition of tincture names; they refer to the order in which the tinctures were first mentioned. "Counterchanged" means that a charge which straddles a line of division is tinctured of the same tinctures as the divided field, reversed (see Behnsdorf arms pictured above).

But as to the rigid formulae of blazoning, John Brooke-Little, Norroy and Ulster King of Arms, wrote in 1985: "Although there are certain conventions as to how arms shall be blazoned ... many of the supposedly hard and fast rules laid down in heraldic manuals [including those by heralds] are often ignored."[3]

A given coat-of-arms may be drawn in many different ways, all considered equivalent, just as the letter "A" may be printed in many different fonts while still being the same letter. For example, the shape of the escutcheon is almost always immaterial, so long as it is not of an anachronistic variety, a rare exception being the coat of arms of Nunavut, for which a round shield is specified.

Because heraldry developed at a time when English clerks wrote in Anglo-Norman French, many terms in English heraldry are of French origin, as is the practice of placing most adjectives after nouns rather than before.

Several websites (cited below) give illustrations of coats of arms with blazons; these can give the reader a good practical feel for blazons and blazoning.

Tinctures

Tincture is the term used to describe the background of a field. It can be a color, a metal or a fur (i.e. pattern). In a black and white representation of arms (such as in bookplates), colors are represented through the use of hatching or patterns of lines and dots. The list of standard tinctures and their names are listed below.

- Metals
  - Or (gold)
  - Argent (silver)
- Colors
  - gules (red)
  - azure (blue)
  - sable (black)
  - vert (green)
  - purpure (purple)
  - tenné (orange)
  - sanguine (blood red)
- Furs
  - ermine
  - vair
  - potent

Complexity

Full descriptions of shields range in complexity, from a single word to a convoluted series describing compound shields:

- Arms of Brittany: *Ermine*
- *Azure, a bend Or*, over which the families of Scrope and Grosvenor fought a famous legal battle (see Scrope v. Grosvenor and image above).
- Arms of Östergötland, Sweden: *Gules a griffin with dragon wings, tail and tongue rampant Or armed, beaked, langued and membered azure between four roses argent.*
- Arms of Hungary dating from 1867, when part of Austria-Hungary:

Quarterly, I azure, three lions' heads affrontés crowned Or (for Dalmatia); II chequy gules and argent (for Croatia); III azure, a river in fess gules bordered argent thereupon a marten proper beneath a six-pointed star Or (for Slavonia); IV per fess azure and Or, overall a bar gules in the chief a demi-eagle sable displayed addextré of the sun in splendour and senestré of a crescent argent in the base seven towers three and four gules (for Transylvania); enté en point gules, a double-headed eagle proper on a peninsula vert holding a vase pouring water into the sea argent beneath a crown proper with bands azure (for Fiume); overall an escutcheon barry of eight gules and argent impaling gules, on a mount vert a crown Or issuant therefrom a double cross argent (for Hungary).[4]

Arms of Brittany

Arms of Östergötland

Arms of Hungary (1867)

See also

- Flag terminology

References

[1] "blazon, n.". *OED Online.* June 2012. Oxford University Press. http://www.oed.com/view/Entry/20024 (accessed September 10, 2012).
[2] Encyclopaedia Brittanica, 9th ed., vol.11, p.683, "Heraldry"
[3] J P Brooke-Little: *An heraldic alphabet* (new and revised edition), p. 52. Robson Books, London, 1985.

Notes

- Brault, Gerard J. (1997). *Early Blazon: Heraldic Terminology in the Twelfth and Thirteenth Centuries*, (2nd ed.). Woodbridge, UK: The Boydell Press. ISBN 0-85115-711-4.
- Elvin, Charles Norton. (1969). *A Dictionary of Heraldry*. London: Heraldry Today. ISBN 0-900455-00-4.
- Parker, James. *A Glossary of Terms Used in Heraldry*, (2nd ed.). Rutland, VT: Charles E. Tuttle Co. ISBN 0-8048-0715-9.

External links

- Heraldric Dictionary (http://www.rarebooks.nd.edu/digital/heraldry/colors.html#METALS)
- A Heraldic Primer (http://www.sca.org/heraldry/primer/index.html), by Stephen Gold and Timothy Shead, explaining the terminology in detail. URL last accessed 2010-12-27
- Luz Herald (http://www.luz-herald.net/free/): free access to web versions of *Burke's General Armory* (incomplete, 1,500 British surnames) and *Pimbley's Dictionary of Heraldry*, as well as *Blason des familles d'Europe, Grand Armorial Universel* (15,000 European surnames) (in French)
- A Grammar of Blazonry (http://www.sca.org/heraldry/laurel/bruce.html) by Bruce Miller. URL last accessed 2012-08-31.
- "Commonly Known" Heraldic Blazon/Emblazon Knowledge (http://heralds.westkingdom.org/CommonBlazonKnowledge.htm), an SCA page with a lengthy dictionary of blazon terms
- Scottish Ensigns Armorial (http://www.kilmarnock.com/blazon.html), the blazon of the Arms of the Baron of Kilmarnock.
- Public Register of the Canadian Heraldic Authority (http://archive.gg.ca/heraldry/pub-reg/) with many useful official versions of modern coats of arms, searchable online
- Civic Heraldry of England and Wales (http://www.civicheraldry.co.uk), fully searchable with illustrations
- Arms of members of the Heraldry Society of Scotland (http://heraldry-scotland.com/copgal/thumbnails.php?album=7), fully searchable with illustrations of bearings
- Arms of members of the Heraldry Society (England) (http://www.theheraldrysociety.com/resources/members.htm), with illustrations of bearings
- Members' Roll of Arms of the Royal Heraldry Society of Canada (http://www.heraldry.ca/main.php?pg=l1), with illustrations of bearings

Saint-Aubin-Sauges

Saint-Aubin-Sauges	
Country Switzerland **Canton** Neuchâtel **District** Boudry	
Population	2419 (Dec 2011)[1]
- Density	314 /km^2 (813 /sq mi)
Area	km^2 (sq mi)[2]
Elevation	469 m (ft)
Postal code	2024
SFOS number	6414
Surrounded by	Estavayer-le-Lac (FR), Fresens, Gorgier, Montalchez, Vaumarcus
Website	saint-aubin-sauges.ne.ch [3] Profile [4] (French), SFSO statistics [5]

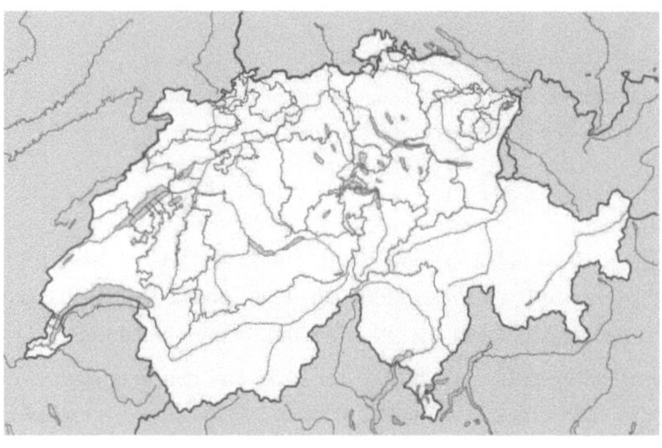

Saint-Aubin-Sauges

Saint-Aubin-Sauges is a municipality in the district of Boudry in the canton of Neuchâtel in Switzerland.

History

Saint-Aubin-Sauges is first mentioned in 1176 as *Sancti Albini*. In 1340 it was mentioned as *villa de Sauges*.[5]

Geography

Port Conty in Saint-Aubin-Sauges on Lake Neuchatel

Saint-Aubin-Sauges has an area, as of 2009, of 7.7 square kilometers (sq mi). Of this area, 3.68 km^2 (sq mi) or 47.7% is used for agricultural purposes, while 2.95 km^2 (sq mi) or 38.3% is forested. Of the rest of the land, 1.04 km^2 (sq mi) or 13.5% is settled (buildings or roads) and 0.02 km^2 (acres) or 0.3% is unproductive land.[6]

Of the built up area, housing and buildings made up 7.8% and transportation infrastructure made up 3.5%. Out of the forested land, 35.0% of the total land area is heavily forested and 3.2% is covered with orchards or small clusters of trees. Of the agricultural land, 16.9% is used for growing crops and 13.6% is pastures, while 3.9% is used for orchards or vine crops and 13.4% is used for alpine pastures.[6]

The municipality is located in the Boudry district, in the La Béroche region. It stretched from the shores of Lake Neuchatel to an elevation of 1436 m (ft). It consists of the villages of Saint-Aubin and Sauges, which merged in 1888.[7]

Coat of arms

The blazon of the municipal coat of arms is *Impalled, per pale Argent and Gules a Rose counterchanged, and per pale Argent three bars Gules and Gules a Rose counterchanged.*[8]

Demographics

Saint-Aubin-Sauges has a population (as of December 2011) of 2419.[1] As of 2008, 21.2% of the population are resident foreign nationals.[9] Over the last 10 years (2000–2010) the population has changed at a rate of 0.8%. It has changed at a rate of 0.6% due to migration and at a rate of 1% due to births and deaths.[10]

Most of the population (as of 2000) speaks French (2,054 or 84.7%) as their first language, Portuguese is the second most common (106 or 4.4%) and German is the third (105 or 4.3%). There are 90 people who speak Italian and 1 person who speaks Romansh.[11]

As of 2008, the population was 47.7% male and 52.3% female. The population was made up of 896 Swiss men (36.9% of the population) and 261 (10.8%) non-Swiss men. There were 1,054 Swiss women (43.4%) and 215 (8.9%) non-Swiss women.[12] Of the population in the municipality, 587 or about 24.2% were born in Saint-Aubin-Sauges and lived there in 2000. There were 603 or 24.9% who were born in the same canton, while 592 or 24.4% were born somewhere else in Switzerland, and 582 or 24.0% were born outside of Switzerland.[11]

As of 2000, children and teenagers (0–19 years old) make up 23.3% of the population, while adults (20–64 years old) make up 59.7% and seniors (over 64 years old) make up 17%.[10]

As of 2000, there were 939 people who were single and never married in the municipality. There were 1,169 married individuals, 166 widows or widowers and 152 individuals who are divorced.[11]

As of 2000, there were 1,008 private households in the municipality, and an average of 2.3 persons per household.[10] There were 336 households that consist of only one person and 65 households with five or more people. In 2000, a total of 986 apartments (85.1% of the total) were permanently occupied, while 138 apartments (11.9%) were seasonally occupied and 35 apartments (3.0%) were empty.[13] As of 2009, the construction rate of new housing units was 0.4 new units per 1000 residents.[10] The vacancy rate for the municipality, in 2010, was 0.81%.[10]

The historical population is given in the following chart:[5] [14]

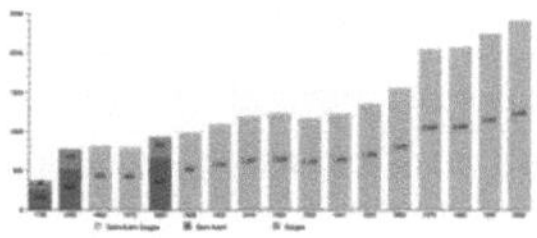

Heritage sites of national significance

The Le Béroche, a Gallo-Roman settlement, and the neolithic settlement at Port Conty / Tivoli are listed as Swiss heritage site of national significance.[15]

World heritage site

It is home to the Port-Conty prehistoric pile-dwelling (or stilt house) settlement that is part of the Prehistoric Pile dwellings around the Alps UNESCO World Heritage Site.[16]

Port-Conty has two neolithic settlements. The first is a Late Cortaillod or Port-Conty type Cortaillod village. One piece of timber from this site has been dated to 3574. The second village is a Late Horgen village. Five of the piles are from 3160 and 3159, and two piles are from 3064 and 3062. The site was discovered in 1860 by F. Troyon, who originally thought it was a Bronze Age village. Further excavations in the late 19th and early 20th century found that it was a neolithic site. In 1929-1932, P. Vouga dug a long (55 m [Category:Pages with bad rounding precision ft]) 1 to 1.5 m (**unknown operator: u'['** to **unknown operator: u'['** ft) deep trench to examine a single layer. He found three archaeological horizons which were from the Middle Neolithic, the Late Neolithic and the end of the Late Neolithic. Based on his study, it appears that the site was covered by an artificial mound which was 20 m × 10 m (**unknown operator: u'['** ft × **unknown operator: u'['** ft) and about 70 cm () deep. It was built in two stages, with a layer of branches between the layers.[17]

La Béroche site

Port Conty

Twin Town

Saint-Aubin-Sauges is twinned with the town of Untersiggenthal, Solothurn.[18]

Politics

In the 2007 federal election the most popular party was the SP which received 27.34% of the vote. The next three most popular parties were the LPS Party (18.64%), the SVP (18.46%) and the FDP (10.68%). In the federal election, a total of 779 votes were cast, and the voter turnout was 50.7%.[19]

Economy

As of 2010, Saint-Aubin-Sauges had an unemployment rate of 5.8%. As of 2008, there were 31 people employed in the primary economic sector and about 13 businesses involved in this sector. 295 people were employed in the secondary sector and there were 37 businesses in this sector. 505 people were employed in the tertiary sector, with 78 businesses in this sector.[10] There were 1,176 residents of the municipality who were employed in some capacity, of which females made up 44.0% of the workforce.

In 2008 the total number of full-time equivalent jobs was 691. The number of jobs in the primary sector was 26, of which 21 were in agriculture, 2 were in forestry or lumber production and 3 were in fishing or fisheries. The number of jobs in the secondary sector was 281 of which 152 or (54.1%) were in manufacturing and 127 (45.2%) were in construction. The number of jobs in the tertiary sector was 384. In the tertiary sector; 89 or 23.2% were in wholesale or retail sales or the repair of motor vehicles, 10 or 2.6% were in the movement and storage of goods, 17 or 4.4% were in a hotel or restaurant, 1 was in the information industry, 4 or 1.0% were the insurance or financial industry, 29 or 7.6% were technical professionals or scientists, 13 or 3.4% were in education and 192 or 50.0% were in health care.[20]

In 2000, there were 418 workers who commuted into the municipality and 760 workers who commuted away. The municipality is a net exporter of workers, with about 1.8 workers leaving the municipality for every one entering. About 1.9% of the workforce coming into Saint-Aubin-Sauges are coming from outside Switzerland.[21] Of the working population, 10.1% used public transportation to get to work, and 63.4% used a private car.[10]

Religion

From the 2000 census, 793 or 32.7% were Roman Catholic, while 860 or 35.4% belonged to the Swiss Reformed Church. Of the rest of the population, there were 31 members of an Orthodox church (or about 1.28% of the population), there was 1 individual who belongs to the Christian Catholic Church, and there were 211 individuals (or about 8.70% of the population) who belonged to another Christian church. There were 3 individuals (or about 0.12% of the population) who were Jewish, and 23 (or about 0.95% of the population) who were Islamic. There were 4 individuals who were Buddhist and 6 individuals who belonged to another church. 507 (or about 20.90% of the population) belonged to no church, are agnostic or atheist, and 92 individuals (or about 3.79% of the population) did not answer the question.[11]

Education

In Saint-Aubin-Sauges about 874 or (36.0%) of the population have completed non-mandatory upper secondary education, and 315 or (13.0%) have completed additional higher education (either university or a *Fachhochschule*). Of the 315 who completed tertiary schooling, 50.5% were Swiss men, 32.4% were Swiss women, 11.1% were non-Swiss men and 6.0% were non-Swiss women.[11]

In the canton of Neuchâtel most municipalities provide two years of non-mandatory kindergarten, followed by five years of mandatory primary education. The next four years of mandatory secondary education is provided at thirteen larger secondary schools, which many students travel out of their home municipality to attend.[22] The kindergarten in Saint-Aubin-Sauges is combined with Vaumarcus. During the 2010-11 school year, there were 3 kindergarten classes with a total of 50 students between the municipalities. In the same year, there were 7 primary classes with a total of 133 students.[23]

As of 2000, there were 26 students in Saint-Aubin-Sauges who came from another municipality, while 192 residents attended schools outside the municipality.[21]

References

[1] Swiss Federal Statistics Office – STAT-TAB (http://www.pxweb.bfs.admin.ch/Dialog/varval.asp?ma=px-d-01-2A01&ti=Ständige+und+Nichtständige+Wohnbevölkerung+nach+Region,+Geschlecht,+Nationalität+und++Alter&path=../Database/German_01 - Bevölkerung/01.2 - Bevölkerungsstand und -bewegung/&lang=1&prod=01&openChild=true&secprod=2) *Ständige und Nichtständige Wohnbevölkerung nach Region, Geschlecht, Nationalität und Alter* **(German)** accessed 3 October 2012

[2] Arealstatistik Standard - Gemeindedaten nach 4 Hauptbereichen (http://www.bfs.admin.ch/bfs/portal/de/index/themen/02/03/blank/data/gemeindedaten.html)

[3] http://saint-aubin-sauges.ne.ch

[4] http://www.ne.ch/neat/documents/statistiques/Fiches_4829/wportraits_Files/NE6414.pdf

[5] *Saint-Aubin-Sauges* in German (http://www.hls-dhs-dss.ch/textes/d/D2836.php), French (http://www.hls-dhs-dss.ch/textes/f/F2836.php) and Italian (http://www.hls-dhs-dss.ch/textes/i/I2836.php) in the online *Historical Dictionary of Switzerland*.

[6] Swiss Federal Statistical Office-Land Use Statistics (http://www.bfs.admin.ch/bfs/portal/de/index/themen/02/03/blank/data/gemeindedaten.html) 2009 data **(German)** accessed 25 March 2010

[7] Amtliches Gemeindeverzeichnis der Schweiz (http://www.bfs.admin.ch/bfs/portal/de/index/infothek/nomenklaturen/blank/blank/gem_liste/03.html) published by the Swiss Federal Statistical Office **(German)** accessed 19 July 2011

[8] Flags of the World.com (http://www.crwflags.com/fotw/flags/ch-ne014.html) accessed 20-October-2011

[9] Swiss Federal Statistical Office - Superweb database - Gemeinde Statistics 1981-2008 (http://www.bfs.admin.ch/bfs/portal/de/index/infothek/onlinedb/superweb/login.html) **(German)** accessed 19 June 2010

[10] Swiss Federal Statistical Office (http://www.bfs.admin.ch/bfs/portal/en/index/regionen/02/key.html) accessed 20-October-2011

[11] STAT-TAB Datenwürfel für Thema 40.3 - 2000 (http://www.pxweb.bfs.admin.ch/Database/German_40 - EidgenÃ¶ssische VolkszÃ¤hlung/40.3 - 2000/40.3 - 2000.asp?lang=1&prod=40&secprod=3&openChild=true) **(German)** accessed 2 February 2011

[12] Canton of Neuchatel Statistics (http://www.ne.ch/neat/site/jsp/rubrique/rubrique.jsp?StyleType=bleu&DocId=40383), *République et canton de Neuchâtel - Recensement annuel de la population* **(German)** accessed 13 October 2011

[13] Swiss Federal Statistical Office STAT-TAB - Datenwürfel für Thema 09.2 - Gebäude und Wohnungen (http://www.pxweb.bfs.admin.ch/Database/German_09 - Bau- und Wohnungswesen/09.2 - GebÃ¤ude und Wohnungen/09.2 - GebÃ¤ude und Wohnungen.asp?lang=1&prod=09&secprod=2&openChild=true) **(German)** accessed 28 January 2011

[14] Swiss Federal Statistical Office STAT-TAB Bevölkerungsentwicklung nach Region, 1850-2000 (http://www.pxweb.bfs.admin.ch/Database/German_40 - EidgenÃ¶ssische VolkszÃ¤hlung/40.4 - 1850-2000/40.4 - 1850-2000.asp?lang=1&prod=40&secprod=4&openChild=true) **(German)** accessed 29 January 2011

[16] UNESCO World Heritage Site - Prehistoric Pile dwellings around the Alps (http://whc.unesco.org/en/list/1363/)

[17] palafittes.org UNESCO nomination files-Volume II: Id-files of the component parts of the serial, Sites Switzerland (2) (http://www.palafittes.org/uploads/media/Volume_II_CH_P_96-207.pdf) accessed20-October-2011

[18] Conseil des Communes et Regions d'Europe (http://www.asccre.ch/N200/neuchatel-neuenburg-neuchatel/?M=222) **(French)** accessed 27 April 2011

[19] Swiss Federal Statistical Office, *Nationalratswahlen 2007: Stärke der Parteien und Wahlbeteiligung, nach Gemeinden/Bezirk/Canton* (http://www.bfs.admin.ch/bfs/portal/de/index/themen/17/02/blank/data/04/03.html) **(German)** accessed 28 May 2010

[20] Swiss Federal Statistical Office STAT-TAB Betriebszählung: Arbeitsstätten nach Gemeinde und NOGA 2008 (Abschnitte), Sektoren 1-3 (http://www.pxweb.bfs.admin.ch/Database/German_06 - Industrie und Dienstleistungen/06.2 - Unternehmen/06.2 - Unternehmen.asp?lang=1&prod=06&secprod=2&openChild=true) **(German)** accessed 28 January 2011

[21] Swiss Federal Statistical Office - Statweb (http://www.media-stat.admin.ch/stat/pendler/pop.php) **(German)** accessed 24 June 2010

[22] EDK/CDIP/IDES (2010). Kantonale Schulstrukturen in der Schweiz und im Fürstentum Liechtenstein / Structures Scolaires Cantonales en Suisse et Dans la Principauté du Liechtenstein (http://edudoc.ch/record/35128/files/Schulsystem_alle.pdf) (Report). . Retrieved 24 June 2010.

[23] Statistical Department of the Canton of Neuchâtel (http://www.ne.ch/neat/site/jsp/rubrique/rubrique.jsp?CatId=9939) Mémento de l'année scolaire 2010/2011 (French) accessed 17 October 2011

Provence,_Switzerland

Provence

Country	Switzerland
Canton	Vaud
District	Jura-Nord Vaudois

Population	345 (Dec 2011)[1]
- Density	11 /km^2 (28 /sq mi)
Area	km^2 (sq mi)[2]
Elevation	778 m (ft)
Postal code	1428
SFOS number	5566
Surrounded by	Bonvillars, Concise, Couvet (NE), Fontanezier, Fresens (NE), Gorgier (NE), Montalchez (NE), Môtiers (NE), Mutrux, Romairon, Travers (NE)
Website	Provence.ch [3] Profile [4], SFSO statistics [5]

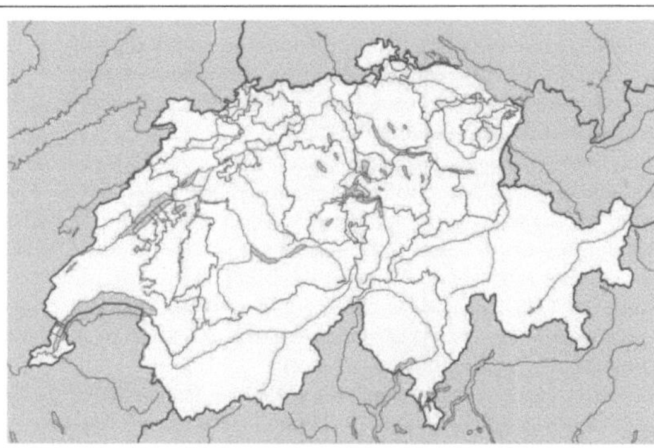

Provence

Provence is a municipality in the district of Jura-Nord Vaudois in the canton of Vaud in Switzerland.

History

Provence is first mentioned in 1340 as *Provency*.[5]

Geography

Provence has an area, as of 2009, of 31.84 square kilometers (sq mi). Of this area, 17.36 km^2 (sq mi) or 54.5% is used for agricultural purposes, while 13.74 km^2 (sq mi) or 43.2% is forested. Of the rest of the land, 0.7 km^2 (sq mi) or 2.2% is settled (buildings or roads) and 0.06 km^2 (acres) or 0.2% is unproductive land.[6]

Of the built up area, housing and buildings made up 0.7% and transportation infrastructure made up 1.5%. Out of the forested land, 36.1% of the total land area is heavily forested and 7.0% is covered with orchards or small clusters of trees. Of the agricultural land, 3.2% is used for growing crops and 12.4% is pastures and 38.6% is used for alpine pastures.[6]

The municipality was part of the Grandson District until it was dissolved on 31 August 2006, and Provence became part of the new district of Jura-Nord Vaudois.[7]

The municipality is located on the border of the Val-de-Travers and stretches from the foot of Mont Aubert to Creux-du-Van. It consists of the village of Provence, the hamlet of Les Prises, the housing development of Montagne Devant and the La Nouvelle Censière area.

Coat of arms

The blazon of the municipal coat of arms is *Pally of Six Argent and Azure, a Buck Gules crowned with a Latin cross emanating Or.*[8]

Demographics

Provence has a population (as of December 2011) of 345.[1] As of 2008, 4.9% of the population are resident foreign nationals.[9] Over the last 10 years (1999–2009) the population has changed at a rate of 5.6%. It has changed at a rate of 8.3% due to migration and at a rate of -2.7% due to births and deaths.[10]

Most of the population (as of 2000) speaks French (354 or 96.2%), with German being second most common (11 or 3.0%) and Italian being third (1 or 0.3%).[11]

Of the population in the municipality 154 or about 41.8% were born in Provence and lived there in 2000. There were 57 or 15.5% who were born in the same canton, while 96 or 26.1% were born somewhere else in Switzerland, and 25 or 6.8% were born outside of Switzerland.[11]

In 2008 there were 3 live births to Swiss citizens and were 2 deaths of Swiss citizens. Ignoring immigration and emigration, the population of Swiss citizens increased by 1 while the foreign population remained the same. There was 1 Swiss man and 1 Swiss woman who emigrated from Switzerland. At the same time, there was 2 non-Swiss women who immigrated from another country to Switzerland. The total Swiss population change in 2008 (from all sources, including moves across municipal borders) was a decrease of 2 and the non-Swiss population increased by 1 people. This represents a population growth rate of -0.3%.[9]

The age distribution, as of 2009, in Provence is; 29 children or 8.1% of the population are between 0 and 9 years old and 54 teenagers or 15.1% are between 10 and 19. Of the adult population, 34 people or 9.5% of the population are between 20 and 29 years old. 46 people or 12.8% are between 30 and 39, 60 people or 16.8% are between 40 and 49, and 41 people or 11.5% are between 50 and 59. The senior population distribution is 45 people or 12.6% of the population are between 60 and 69 years old, 30 people or 8.4% are between 70 and 79, there are 18 people or 5.0% who are between 80 and 89, and there is 1 person who is 90 and older.[12]

As of 2000, there were 165 people who were single and never married in the municipality. There were 154 married individuals, 27 widows or widowers and 22 individuals who are divorced.[11]

As of 2000, there were 144 private households in the municipality, and an average of 2.3 persons per household.[10] There were 48 households that consist of only one person and 13 households with five or more people. Out of a total of 148 households that answered this question, 32.4% were households made up of just one person and there was 1 adult who lived with their parents. Of the rest of the households, there are 40 married couples without children, 43 married couples with children There were 7 single parents with a child or children. There were 5 households that were made up of unrelated people and 4 households that were made up of some sort of institution or another collective housing.[11]

In 2000 there were 51 single family homes (or 40.2% of the total) out of a total of 127 inhabited buildings. There were 21 multi-family buildings (16.5%), along with 44 multi-purpose buildings that were mostly used for housing (34.6%) and 11 other use buildings (commercial or industrial) that also had some housing (8.7%). Of the single family homes 35 were built before 1919, while 5 were built between 1990 and 2000. The most multi-family homes (16) were built before 1919 and the next most (2) were built between 1961 and 1970.[13]

In 2000 there were 165 apartments in the municipality. The most common apartment size was 3 rooms of which there were 46. There were 6 single room apartments and 44 apartments with five or more rooms. Of these apartments, a total of 127 apartments (77.0% of the total) were permanently occupied, while 29 apartments (17.6%) were seasonally occupied and 9 apartments (5.5%) were empty.[13] As of 2009, the construction rate of new housing units was 0 new units per 1000 residents.[10] The vacancy rate for the municipality, in 2010, was 0.58%.[10]

The historical population is given in the following chart:[5] [14]

Heritage sites of national significance

The Chalet D'Alpage La Redalle is listed as a Swiss heritage site of national significance. The entire village of Provence is part of the Inventory of Swiss Heritage Sites.[15]

Politics

In the 2007 federal election the most popular party was the SP which received 30.89% of the vote. The next three most popular parties were the SVP (29.84%), the Green Party (17.43%) and the FDP (7.07%). In the federal election, a total of 100 votes were cast, and the voter turnout was 40.0%.[16]

Economy

As of 2010, Provence had an unemployment rate of 1.5%. As of 2008, there were 62 people employed in the primary economic sector and about 22 businesses involved in this sector. 11 people were employed in the secondary sector and there were 4 businesses in this sector. 47 people were employed in the tertiary sector, with 9 businesses in this sector.[10] There were 163 residents of the municipality who were employed in some capacity, of which females made up 41.7% of the workforce.

In 2008 the total number of full-time equivalent jobs was 94. The number of jobs in the primary sector was 49, of which 46 were in agriculture and 3 were in forestry or lumber production. The number of jobs in the secondary sector was 8, all of which were in manufacturing. The number of jobs in the tertiary sector was 37. In the tertiary sector; 5 or 13.5% were in wholesale or retail sales or the repair of motor vehicles, 8 or 21.6% were in a hotel or restaurant, 1 was in the information industry, 3 or 8.1% were in education and 21 or 56.8% were in health care.[17]

In 2000, there were 31 workers who commuted into the municipality and 80 workers who commuted away. The municipality is a net exporter of workers, with about 2.6 workers leaving the municipality for every one entering.[18] Of the working population, 7.4% used public transportation to get to work, and 50.9% used a private car.[10]

Religion

From the 2000 census, 32 or 8.7% were Roman Catholic, while 243 or 66.0% belonged to the Swiss Reformed Church. Of the rest of the population, there was 1 member of an Orthodox church, and there were 12 individuals (or about 3.26% of the population) who belonged to another Christian church. 56 (or about 15.22% of the population) belonged to no church, are agnostic or atheist, and 30 individuals (or about 8.15% of the population) did not answer the question.[11]

Education

In Provence about 97 or (26.4%) of the population have completed non-mandatory upper secondary education, and 30 or (8.2%) have completed additional higher education (either university or a *Fachhochschule*). Of the 30 who completed tertiary schooling, 53.3% were Swiss men, 36.7% were Swiss women.[11]

In the 2009/2010 school year there were a total of 44 students in the Provence school district. In the Vaud cantonal school system, two years of non-obligatory pre-school are provided by the political districts.[19] During the school year, the political district provided pre-school care for a total of 578 children of which 359 children (62.1%) received subsidized pre-school care. The canton's primary school program requires students to attend for four years. There were 19 students in the municipal primary school program. The obligatory lower secondary school program lasts for six years and there were 22 students in those schools. There were also 3 students who were home schooled or attended another non-traditional school.[20]

As of 2000, there were 11 students in Provence who came from another municipality, while 32 residents attended schools outside the municipality.[18]

References

[1] Swiss Federal Statistics Office – STAT-TAB (http://www.pxweb.bfs.admin.ch/Dialog/varval.asp?ma=px-d-01-2A01&ti=Ständige+und+Nichtständige+Wohnbevölkerung+nach+Region,+Geschlecht,+Nationalität+und++Alter&path=../Database/German_01 - Bevölkerung/01.2 - Bevölkerungsstand und -bewegung/&lang=1&prod=01&openChild=true&secprod=2) *Ständige und Nichtständige Wohnbevölkerung nach Region, Geschlecht, Nationalität und Alter* (German) accessed 3 October 2012

[2] Arealstatistik Standard - Gemeindedaten nach 4 Hauptbereichen (http://www.bfs.admin.ch/bfs/portal/de/index/themen/02/03/blank/data/gemeindedaten.html)

[3] http://Provence.ch/

[4] http://www.ucv.ch/communal/Co_Ucv.asp?NumStr=09.20&NoOFS=5566&Lng=E

[5] *Provence* in German (http://www.hls-dhs-dss.ch/textes/d/D2398.php), French (http://www.hls-dhs-dss.ch/textes/f/F2398.php) and Italian (http://www.hls-dhs-dss.ch/textes/i/I2398.php) in the online *Historical Dictionary of Switzerland.*

[6] Swiss Federal Statistical Office-Land Use Statistics (http://www.bfs.admin.ch/bfs/portal/de/index/themen/02/03/blank/data/gemeindedaten.html) 2009 data (German) accessed 25 March 2010

[7] Nomenklaturen – Amtliches Gemeindeverzeichnis der Schweiz (http://www.bfs.admin.ch/bfs/portal/de/index/infothek/nomenklaturen/blank/blank/gem_liste/02.html) (German) accessed 4 April 2011

[8] Flags of the World.com (http://www.crwflags.com/fotw/flags/ch-vd124.html) accessed 9 June 2011

[9] Swiss Federal Statistical Office - Superweb database - Gemeinde Statistics 1981-2008 (http://www.bfs.admin.ch/bfs/portal/de/index/infothek/onlinedb/superweb/login.html) (German) accessed 19 June 2010

[10] Swiss Federal Statistical Office (http://www.bfs.admin.ch/bfs/portal/en/index/regionen/02/key.html) accessed 9 June 2011

[11] STAT-TAB Datenwürfel für Thema 40.3 - 2000 (http://www.pxweb.bfs.admin.ch/Database/German_40 - EidgenÃ¶ssische VolkszÃ¤hlung/40.3 - 2000/40.3 - 2000.asp?lang=1&prod=40&secprod=3&openChild=true) (German) accessed 2 February 2011

[12] Canton of Vaud Statistical Office (http://www.scris.vd.ch/Default.aspx?DomID=2016) (French) accessed 29 April 2011

[13] Swiss Federal Statistical Office STAT-TAB - Datenwürfel für Thema 09.2 - Gebäude und Wohnungen (http://www.pxweb.bfs.admin.ch/Database/German_09 - Bau- und Wohnungswesen/09.2 - GebÃ¤ude und Wohnungen/09.2 - GebÃ¤ude und Wohnungen.asp?lang=1&prod=09&secprod=2&openChild=true) (German) accessed 28 January 2011

[14] Swiss Federal Statistical Office STAT-TAB Bevölkerungsentwicklung nach Region, 1850-2000 (http://www.pxweb.bfs.admin.ch/Database/German_40 - EidgenÃ¶ssische VolkszÃ¤hlung/40.4 - 1850-2000/40.4 - 1850-2000.asp?lang=1&prod=40&secprod=4&openChild=true) (German) accessed 29 January 2011

[16] Swiss Federal Statistical Office, *Nationalratswahlen 2007: Stärke der Parteien und Wahlbeteiligung, nach Gemeinden/Bezirk/Canton* (http://www.bfs.admin.ch/bfs/portal/de/index/themen/17/02/blank/data/04/03.html) (German) accessed 28 May 2010

[17] Swiss Federal Statistical Office STAT-TAB Betriebszählung: Arbeitsstätten nach Gemeinde und NOGA 2008 (Abschnitte), Sektoren 1-3 (http://www.pxweb.bfs.admin.ch/Database/German_06 - Industrie und Dienstleistungen/06.2 - Unternehmen/06.2 - Unternehmen.asp?lang=1&prod=06&secprod=2&openChild=true) (German) accessed 28 January 2011

[18] Swiss Federal Statistical Office - Statweb (http://www.media-stat.admin.ch/stat/pendler/pop.php) (German) accessed 24 June 2010

[19] Organigramme de l'école vaudoise, année scolaire 2009-2010 (http://www.scris.vd.ch/Default.aspx?DomId=1990) (French) accessed 2 May 2011

[20] Canton of Vaud Statistical Office - Scol. obligatoire/filières de transition (http://www.scris.vd.ch/Default.aspx?DomID=2403) (French) accessed 2 May 2011

Fresens

Fresens	
Country Switzerland **Canton** Neuchâtel **District** Boudry	
Population	222 (Dec 2011)[1]
- Density	141 /km^2 (364 /sq mi)
Area	km^2 (sq mi)[2]
Elevation	620 m (ft)
Postal code	2027
SFOS number	6409
Surrounded by	Montalchez, Mutrux (VD), Provence (VD), Saint-Aubin-Sauges, Vaumarcus
Website	Fresens.ch [3] Profile [4] **(French)**, SFSO statistics [5]

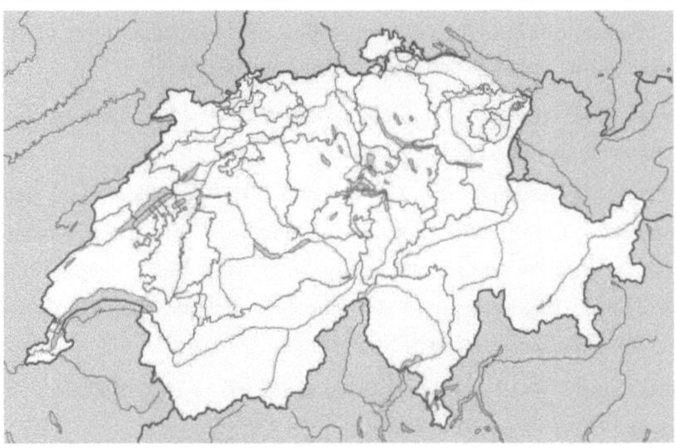

Fresens

Fresens is a municipality in the district of Boudry in the canton of Neuchâtel in Switzerland.

History

Fresens is first mentioned in 1265 as *Hiremanz dit de Fressen*.[5]

Geography

Fresens has an area, as of 2009, of 1.6 square kilometers (sq mi). Of this area, 1.32 km^2 (sq mi) or 83.5% is used for agricultural purposes, while 0.12 km^2 (sq mi) or 7.6% is forested. Of the rest of the land, 0.15 km^2 (acres) or 9.5% is settled (buildings or roads).[6]

Of the built up area, housing and buildings made up 5.1% and transportation infrastructure made up 3.8%. Out of the forested land, 5.7% of the total land area is heavily forested and 1.9% is covered with orchards or small clusters of trees. Of the agricultural land, 64.6% is used for growing crops and 13.3% is pastures, while 5.7% is used for orchards or vine crops.[6]

The municipality is located in the Boudry district.

Coat of arms

The blazon of the municipal coat of arms is *Per fess Gules and Argent, a Rose counterchanged.*[7]

Demographics

Fresens has a population (as of December 2011) of 222.[1] As of 2008, 2.4% of the population are resident foreign nationals.[8] Over the last 10 years (2000–2010) the population has changed at a rate of 10.6%. It has changed at a rate of 2.1% due to migration and at a rate of 7.4% due to births and deaths.[9]

Most of the population (as of 2000) speaks French (206 or 94.5%) as their first language, German is the second most common (8 or 3.7%) and Italian is the third (1 or 0.5%).[10]

As of 2008, the population was 49.8% male and 50.2% female. The population was made up of 105 Swiss men (48.8% of the population) and 2 (0.9%) non-Swiss men. There were 107 Swiss women (49.8%) and 1 (0.5%) non-Swiss women.[11] Of the population in the municipality, 78 or about 35.8% were born in Fresens and lived there in 2000. There were 54 or 24.8% who were born in the same canton, while 62 or 28.4% were born somewhere else in Switzerland, and 15 or 6.9% were born outside of Switzerland.[10]

As of 2000, children and teenagers (0–19 years old) make up 22.9% of the population, while adults (20–64 years old) make up 48.2% and seniors (over 64 years old) make up 28.9%.[9]

As of 2000, there were 85 people who were single and never married in the municipality. There were 98 married individuals, 24 widows or widowers and 11 individuals who are divorced.[10]

As of 2000, there were 62 private households in the municipality, and an average of 2.8 persons per household.[9] There were 10 households that consist of only one person and 9 households with five or more people. In 2000, a total of 60 apartments (83.3% of the total) were permanently occupied, while 6 apartments (8.3%) were seasonally occupied and 6 apartments (8.3%) were empty.[12] As of 2009, the construction rate of new housing units was 4.8 new units per 1000 residents.[9]

The historical population is given in the following chart:[5] [13]

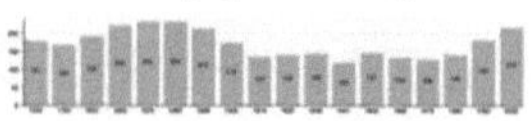

Politics

In the 2007 federal election the most popular party was the SVP which received 32.61% of the vote. The next three most popular parties were the SP (18.86%), the FDP (12.77%) and the Green Party (11.98%). In the federal election, a total of 104 votes were cast, and the voter turnout was 65.8%.[14]

Economy

As of 2010, Fresens had an unemployment rate of 5%. As of 2008, there were 24 people employed in the primary economic sector and about 9 businesses involved in this sector. 78 people were employed in the secondary sector and there were 5 businesses in this sector. 60 people were employed in the tertiary sector, with 3 businesses in this sector.[9] There were 94 residents of the municipality who were employed in some capacity, of which females made up 37.2% of the workforce.

In 2008 the total number of full-time equivalent jobs was 134. The number of jobs in the primary sector was 18, all of which were in agriculture. The number of jobs in the secondary sector was 77 of which 26 or (33.8%) were in manufacturing and 51 (56.2%) were in construction. The number of jobs in the tertiary sector was 39. In the tertiary sector; 2 or 5.1% were in education and 37 or 94.9% were in health care.[15]

In 2000, there were 100 workers who commuted into the municipality and 49 workers who commuted away. The municipality is a net importer of workers, with about 2.0 workers entering the municipality for every one leaving. About 13.0% of the workforce coming into Fresens are coming from outside Switzerland.[16] Of the working population, 5.3% used public transportation to get to work, and 46.8% used a private car.[9]

Religion

From the 2000 census, 8 or 3.7% were Roman Catholic, while 99 or 45.4% belonged to the Swiss Reformed Church. Of the rest of the population, there were 2 members of an Orthodox church (or about 0.92% of the population), and there were 142 individuals (or about 65.14% of the population) who belonged to another Christian church. There were 2 individuals (or about 0.92% of the population) who were Jewish, and there was 1 individual who was Islamic. There were 1 individual who belonged to another church. 25 (or about 11.47% of the population) belonged to no church, are agnostic or atheist, and 9 individuals (or about 4.13% of the population) did not answer the question.[10]

Education

In Fresens about 91 or (41.7%) of the population have completed non-mandatory upper secondary education, and 25 or (11.5%) have completed additional higher education (either university or a *Fachhochschule*). Of the 25 who completed tertiary schooling, 64.0% were Swiss men, 36.0% were Swiss women.[10]

In the canton of Neuchâtel most municipalities provide two years of non-mandatory kindergarten, followed by five years of mandatory primary education. The next four years of mandatory secondary education is provided at thirteen larger secondary schools, which many students travel out of their home municipality to attend.[17] Both the kindergarten and the primary school are combined with Montalchez. During the 2010-11 school year, there was one kindergarten class with a total of 23 students between the municipalities. In the same year, there was one primary class with a total of 16 students.[18]

As of 2000, there were 2 students in Fresens who came from another municipality, while 22 residents attended schools outside the municipality.[16]

References

[1] Swiss Federal Statistics Office – STAT-TAB (http://www.pxweb.bfs.admin.ch/Dialog/varval.asp?ma=px-d-01-2A01&ti=Ständige+und+Nichtständige+Wohnbevölkerung+nach+Region,+Geschlecht,+Nationalität+und++Alter&path=../Database/German_01 - Bevölkerung/01.2 - Bevölkerungsstand und -bewegung/&lang=1&prod=01&openChild=true&secprod=2) *Ständige und Nichtständige Wohnbevölkerung nach Region, Geschlecht, Nationalität und Alter* **(German)** accessed 3 October 2012

[2] Arealstatistik Standard - Gemeindedaten nach 4 Hauptbereichen (http://www.bfs.admin.ch/bfs/portal/de/index/themen/02/03/blank/data/gemeindedaten.html)

[3] http://Fresens.ch/

[4] http://www.ne.ch/neat/documents/statistiques/Fiches_4829/wportraits_Files/NE6409.pdf

[5] *Fresens* in German (http://www.hls-dhs-dss.ch/textes/d/D2831.php), French (http://www.hls-dhs-dss.ch/textes/f/F2831.php) and Italian (http://www.hls-dhs-dss.ch/textes/i/I2831.php) in the online *Historical Dictionary of Switzerland*.

[6] Swiss Federal Statistical Office-Land Use Statistics (http://www.bfs.admin.ch/bfs/portal/de/index/themen/02/03/blank/data/gemeindedaten.html) 2009 data **(German)** accessed 25 March 2010

[7] Flags of the World.com (http://www.crwflags.com/fotw/flags/ch-ne009.html) accessed 20-October-2011

[8] Swiss Federal Statistical Office - Superweb database - Gemeinde Statistics 1981-2008 (http://www.bfs.admin.ch/bfs/portal/de/index/infothek/onlinedb/superweb/login.html) **(German)** accessed 19 June 2010

[9] Swiss Federal Statistical Office (http://www.bfs.admin.ch/bfs/portal/en/index/regionen/02/key.html) accessed 20-October-2011

[10] STAT-TAB Datenwürfel für Thema 40.3 - 2000 (http://www.pxweb.bfs.admin.ch/Database/German_40 - EidgenÃ¶ssische VolkszÃ¤hlung/40.3 - 2000/40.3 - 2000.asp?lang=1&prod=40&secprod=3&openChild=true) **(German)** accessed 2 February 2011

[11] Canton of Neuchatel Statistics (http://www.ne.ch/neat/site/jsp/rubrique/rubrique.jsp?StyleType=bleu&DocId=40383), *République et canton de Neuchâtel - Recensement annuel de la population* **(German)** accessed 13 October 2011

[12] Swiss Federal Statistical Office STAT-TAB - Datenwürfel für Thema 09.2 - Gebäude und Wohnungen (http://www.pxweb.bfs.admin.ch/Database/German_09 - Bau- und Wohnungswesen/09.2 - GebÃ¤ude und Wohnungen/09.2 - GebÃ¤ude und Wohnungen.asp?lang=1&prod=09&secprod=2&openChild=true) **(German)** accessed 28 January 2011

[13] Swiss Federal Statistical Office STAT-TAB Bevölkerungsentwicklung nach Region, 1850-2000 (http://www.pxweb.bfs.admin.ch/Database/German_40 - EidgenÃ¶ssische VolkszÃ¤hlung/40.4 - 1850-2000/40.4 - 1850-2000.asp?lang=1&prod=40&secprod=4&openChild=true) **(German)** accessed 29 January 2011

[14] Swiss Federal Statistical Office, *Nationalratswahlen 2007: Stärke der Parteien und Wahlbeteiligung, nach Gemeinden/Bezirk/Canton* (http://www.bfs.admin.ch/bfs/portal/de/index/themen/17/02/blank/data/04/03.html) **(German)** accessed 28 May 2010

[15] Swiss Federal Statistical Office STAT-TAB Betriebszählung: Arbeitsstätten nach Gemeinde und NOGA 2008 (Abschnitte), Sektoren 1-3 (http://www.pxweb.bfs.admin.ch/Database/German_06 - Industrie und Dienstleistungen/06.2 - Unternehmen/06.2 - Unternehmen.asp?lang=1&prod=06&secprod=2&openChild=true) **(German)** accessed 28 January 2011

[16] Swiss Federal Statistical Office - Statweb (http://www.media-stat.admin.ch/stat/pendler/pop.php) **(German)** accessed 24 June 2010

[17] EDK/CDIP/IDES (2010). Kantonale Schulstrukturen in der Schweiz und im Fürstentum Liechtenstein / Structures Scolaires Cantonales en Suisse et Dans la Principauté du Liechtenstein (http://edudoc.ch/record/35128/files/Schulsystem_alle.pdf) (Report). . Retrieved 24 June 2010.

[18] Statistical Department of the Canton of Neuchâtel (http://www.ne.ch/neat/site/jsp/rubrique/rubrique.jsp?CatId=9939) Mémento de l'année scolaire 2010/2011 **(French)** accessed 17 October 2011

External links

- *Fresens* in German (http://www.hls-dhs-dss.ch/textes/d/D2831.php), French (http://www.hls-dhs-dss.ch/textes/f/F2831.php) and Italian (http://www.hls-dhs-dss.ch/textes/i/I2831.php) in the online *Historical Dictionary of Switzerland*.

Article Sources and Contributors

Montalchez *Source*: http://en.wikipedia.org/w/index.php?title=Montalchez *Contributors*: Adlei86, BillC, D6, DidiWeidmann, Sole Soul, Tamfang, Tobyc75, 1 anonymous edits

Communes_of_Switzerland *Source*: http://en.wikipedia.org/w/index.php?title=Communes_of_Switzerland *Contributors*: 159753, 2Neuenburg, Acntx, Amikake3, Andonic, Booksworm, Chanheigeorge, Corti, Crusoe8181, Cusio, D6, Db1944, Dbachmann, Dirrival, Docu, Elipongo, Emerson7, EnOreg, Felix Portier, Frietjes, Geschichte, Gestumblindi, Grutness, Hadal, Haruo, Irmgard, J. 'mach' wust, Jurema Oliveira, LeoNomis, Lugia2453, MBisanz, MK, Megan1967, Melo86, NawlinWiki, Onco p53, PsY.cHo, Reindra, Rich Farmbrough, Roland zh, RubleuleR, Schutz, Sectryan, Sietse Snel, Srleffler, Tobias Conradi, Tobyc75, Ultimate Destiny, Wmahan, Woohookitty, Yellowdesk, 27 anonymous edits

Boudry_District *Source*: http://en.wikipedia.org/w/index.php?title=Boudry_District *Contributors*: Adlei86, BillC, DidiWeidmann, Rigadoun, Saltose, Schwyz, Tobyc75, 1 anonymous edits

Cantons_of_Switzerland *Source*: http://en.wikipedia.org/w/index.php?title=Cantons_of_Switzerland *Contributors*: 05, 159753, 2Neuenburg, 3riol, A bit iffy, A. Parrot, AP1787, Acntx, Ahuskay, Alarob, Aram33, Argasp, Arthena, Bag, Barticus88, Bearcat, Blanche of King's Lynn, Brammen, Briséis, Brittainia, BrownHairedGirl, CUSENZA Mario, Cassowary, Chanheigeorge, Chris j wood, ChrisfromHouston, Chùn-tiàn, Claude girardin, Closedmouth, Cnilep, Cunado19, D6, DDima, Dbachmann, Dcoetzee, Deacon of Pndapetzim, Denelson83, Djinn112, Docu, Dzordzm, EamonnPKeane, EdwinHJ, ElC, Electionworld, Enlil Ninlil, Ericamick, Fan-1967, Filzstift, Frank101, Gazilion, Gestumblindi, Gurch, Haddiscoe, Hashar, Hayden120, Hinote, Hmains, Hroðulf, Hunnjazal, Hvn0413, Indigoeleven, Insilvis, Janburse, JdeJ, Jschwa1, Jusjih, Kahusi, Kerryh r, Kokiri, Korovioff, Lightmouse, Lkmorlan, Longbow4u, Luuva, MadGeographer, Martg76, Mascia Luccova, Master of Puppets, Math Champion, Maxí, Megan1967, MeltBanana, Mistakefinder, MrRadioGuy, Mrt3366, Mschlindwein, Mthwaite, Murtasa, N2e, Nalta, Ngagnebin, Nilfanion, Nohat, Nowhither, Ojw, OwenBlacker, Pascal, Patrick, Paul-L, Piccadilly, Piccolo Modificatore Laborioso, Pne, Postdlf, PsY.cHo, Random user 39849958, Rich Farmbrough, Rjwilmsi, Roke, Rpckvv, Sallygirl1783, Sam Blacketer, Sandstein, Sardanaphalus, Schutz, ScottMainwaring, Sesel, SetRochele, SetRochelewsad, Sherool, SiBr4, Skew-t, Sluzzelin, Smjg, Stephensuleeman, Stevietheman, Tamfang, TenOfAllTrades, Thorsen, Tobias Conradi, Tobyc75, Tonymec, Ultimate Destiny, Uncle Scrooge, Vdiest, West.andrew.g, Woohookitty, Worobiew, Wshun, Zacharie Grossen, Zondor, Žiga, პაატა შ, 103 anonymous edits

Canton_of_Neuchâtel *Source*: http://en.wikipedia.org/w/index.php?title=Canton_of_Neuch%C3%A2tel *Contributors*: 2Neuenburg, 2old bill, Adlei86, Ahoerstemeier, Al Silonov, Alanmoss, Amnon s, Aram33, Bastin, BillC, Charles01, ChrisK, Claude girardin, Cymru123, D6, Deacon of Pndapetzim, Denelson83, Der Eberswalder, Dewritech, Djinn112, Docu, Eurotraveler, EvaristoAugello, Flint and Fire, Funnyhat, Gb1291, Goldfishbutt, HeartofaDog, Hunnjazal, Innotata, Jauhienij, Jnc, KirkEN, Kokiri, Kpjas, Lemondrop47, Longbow4u, MIKHEIL, Macronyx, Marek69, Martpol, N2100, Necronox, Neuenburg, Nortmannus, Numbo3, Olessi, Olivier, Oreo Priest, OwenBlacker, P cristina13, P. S. Burton, PBP, Pedant17, Pjetter, Polarpanda, PsY.cHo, R'n'B, Rich Farmbrough, Rjwilmsi, Rror, Sardanaphalus, Saxbryn, Schwyz, Snowdog, SpeDIt, Squash Racket, TGC55, Thecurran, Tobyc75, Tomas e, Trilobite, Trumpnewtq, Ulf Heinsohn, Walden69, Wikibob, Xyzzyva, ZH8000, Zacharie Grossen, Évangéline, 48 anonymous edits

Coat_of_arms *Source*: http://en.wikipedia.org/w/index.php?title=Coat_of_arms *Contributors*: 123aidan, 1exec1, 28421u2232nfenfcenc, A More Perfect Onion, A little insignificant, A.zstephens, ABF, ATurtle05, Addshore, Adhyatma3010, Aervanath, Aetylus, Agentscott00, Ahoerstemeier, Ahuskay, Aka042, Alansohn, Alexius08, Amateur55, Amjad Abdullah, Andre Engels, Angela, AnonMoos, Anonymous from the 21st century, Antalagy, Apeloverage, Arab Hafez, Araignee, ArnoldReinhold, Artyom, Arwel Parry, Asbestos, Ashish Bakshi, Astrotrain, Attilios, BIL, Bakabaka, BanyanTree, Barneca, Basil Bernar, Beetstra, Bennylin, Betacommand, Bewareofdog, Big Adamsky, BilCat, Bissinger, Blauendorn, Bobo192, BolinhasFofas, Borameer, Brandon5485, Brian0918, Brianga, BryanG, Bubbha, C9Rav9c, CALR, CTZMSC3, Cahk, CambridgeBayWeather, Can't sleep, clown will eat me, Capricorn42, CardinalDan, Care, CaribDigita, Carlsmith, Celtus, Centauri, Charruyego, Checkpoint, ChrisOelmueller, Chrism, Christian75, Chwyatt, Circeus, Clear&Simple, Codex Sinaiticus, CommonsDelinker, Cordless Larry, Correogsk, Craigy144, CristianChirita, Crzrussian, Curb Chain, Custodiansoftime, Czar Brodie, DARTH SIDIOUS 2, DTRY, Danbarnesdavies, Danceswithzerglings, Dancingwombatsrule, Danrok, Darcij, Darth Panda, David Latapie, David Shay, David de Cooman, Dawn Bard, Dbachmann, Dcandeto, Debouncer, Decora, Delirium, Demiurge, Derek Andrews, DeusMP, Dfflick319, Digger845, Discospinster, Dlae, Dogears, Doops, Dr pda, DrHollisCollier, Dthomsen8, Duffman, Dysepsion, ESkog, Eliz81, Ellmist, EncycloPetey, Enviroboy, Enzo Aquarius, Epbr123, Equendil, Eric-Wester, Euchiasmus, Eulyzes, Evadb, Evertype, Everyking, Evil saltine, Evrik, Ezeu, Fanatix, Fengxianjin, FiggyBee, Finavon, Fiyinite7, Flausy, Foxcommunication, FrancoGG, Freakofnurture, FrederickSommer, Freederick, Fritzpoll, Fvasconcellos, Gakusha, Galoubet, Galwhaa, Garryq, GearedBull, Generalseven, Geniac, Geoffrey.landis, George.moug, Gilliam, Gimboid13, Gimmetrow, Gire 3pich2005, Glane23, Glenlarson, Gogo Dodo, Gorovich, Gregmatthews, Group29, Grutness, Gsp8181, Gtstricky, Gustav von Humpelschmumpel, Gustavo Szwedowski de Korwin, Hadžija, Halmstad, HamburgerRadio, Hawjam, Heraldics, Herostratus, Hindustanilanguage, Hmains, Hoss789, Hu12, Huntscorpio, INkubusse, IRP, Iaberis, Iamunknown, Igoldste, Imagecrafting, ImperatorExercitus, Inka 888, Innerche, Ipankonin, Iridescent, Islescape, J.delanoy, J36miles, JFreeman, Jackol, JackyR, Jake Wartenberg, JamesBWatson, Jaykjellybean, Jdforrester, JerryFriedman, Jhamez84, Jiang, Jjojj, Jncraton, John, John254, JohnCD, Johnathan Tsui, Jonjames1986, Joseph Solis in Australia, José Luís Ávila Silveira, Jschnur, Jusdafax, Jyusin, Kahner5150, Kaihsu, Kanguole, Kansoku, Karmosin, Keilana, Kelisi, Kevin96850, Khoikhoi, Kimon, King of Hearts, KingNewbs, Kintetsubuffalo, Kiore, Kittybrewster, Kookyunii, Kool Lat'n SD, Kostisl, Kovacho, Kpgh99, Krylonblue83, Kylu, Kyuko, Lacrimosus, Lahiru k, Lambiam, Lammoth, Larry1972, LeaveSleaves, LiDaobing, LibLord, Linkboyz, Lobsterthermidor, Logan, Lord Brantus, Lord Fitzwarin, Loren.wilton, Lradrama, LtNOWIS, M.K, M0rphzone, MER-C, MacGonagall, Madhero88, Magicuser3, Magioladitis, Magog the Ogre, Man vyi, Marcelivan, Mark, Martykoch, MathHisSci, Mauls, Maycoll F. Vieira, Mb1000, Mellisa Anthony Jones, Mhking, Mich Taylor, Michael Hardy, Michaelsanders, Midgrid, Mike Rosoft, Milford212, Mingwangx, Mon Vier, Mormegil, Mothmolevna, Mouser, Moyogo, Munboy, Muriel Gottrop, Mushroom, MusikAnimal, Mygerardromance, Nadav1, Nakon, Nameneko, Neurolysis, NewEnglandYankee, Nilfanion, Nivix, Nkcs, Noropdoropi, Nsaa, Ntsimp, OHCLOUDSUNFOLD, Octane, Ohconfucius, Orange Suede Sofa, Owned Souls, Oxymoron83, PBS, PDH, Palnatoke, Paul Barlow, Paul venter, Pdcook, Pe-Jo, Persian Poet Gal, Petri Krohn, PhJ, Phaedriel, Philip Trueman, Philipandrew, Pieneman1, Pinethicket, Pinkadelica, Pkhg199, PleaseStand, Pmanderson, Pne, Poco a poco, Pojanji, Pokrajac, Porqin, Possum, Prothonotar, Puddhe, Quite, Qxz, Qyd, RandomStringOfCharacters, RapidR, Razakel19, Rdsmith4, Remuel, Rettetast, Reywas92, Rhurtadon, Ricardocolombia, Richard A McDonald, RichardGonzalesV, Rjstott, Robchurch, Rockhall, Ronhjones, Roux, Royboycrashfan, S, SChockey97, SPhotographer, Sadalmelik, Saga City, Salvadorjo, Sammella99, Sbwoodside, Schockey14, SchreiberBike, SeanR70, Sephiroth storm, Shanel, Silverhorse, SimonP, Simonmetcalf, Sinaloa, Sinhaempire, Sionus, Siroxo, Sistan, Skater, Sketraw, Skull 'n' Femurs, Sky Harbor, SmallPotatoes, Smalljim, Smoon17, Snow Blizzard, Snowmanradio, Snowolf, Sotakeit, SowelBlack, Spencer, Sputnikcccp, Srnec, Stansult, Strike Eagle, Studioce, SupermanReturns, SusikMkr, Sweetloulou, TJ Spyke, Tabletop, Tamfang, Tauex, Tckma, Techman224, Tedder, Teo64x, Terot, Terra Xin, TerraHikaru, Thakurji, The Thing That Should Not Be, Theone00, Thingg, This user has left wikipedia, Thisbites, Thisisborin9, Thomas Blomberg, Thumperward, ThuranX, Tide rolls, Tilstock, Tirthajyoti, Tmopkisn, Tom.k, Tone, Tozzano, Trusilver, Twinsor, TylerDurden1963, Tyrenius, Uhai, UserDoe, Utcursch, VAwebteam, Valentinian, Vanished User 1004, VanishedUserABC, Vasilcho, Vincent Liu, Vrenator, Vuo, Vzbs34, WJBscribe, Wales, Wayne Slam, Whistlesgowhoo, Whpq, Wikipedian2, Wilhelm meis, Witan, Woohookitty, Writerguy, Writtenright, Xanderliptak, Xn4, Yamakiri, Yarp, Yopie, Yworo, Zalgo, Zalhael, Zaahe, Zarthon, Zaxo1994, , , 958 anonymous edits

Blazon *Source*: http://en.wikipedia.org/w/index.php?title=Blazon *Contributors*: 3DS Mike, A. di M., Alansohn, Alarob, Alekjds, Altenmann, AndreasPraefcke, AndrewWatt, Anthony Appleyard, Arz1969, Awien, Balmung0731, Bdesham, Birdoman, Bjankuloski06en, Boven, Carnildo, Dancingwombatsrule, Daniel C. Boyer, David Latapie, Denelson83, Digitalme, Dr pda, Edward Waverley, EncycloPetey, Evertype, Fishal, Fool4jesus, Fred J, Furrykef, Fëaluinix, Galoubet, George Burgess, Gimmetrow, Grutness, Gustav von Humpelschmumpel, Halibutt, Hede2000, Hedgie, Hu12, INVERTED, Indefatigable, Jane023, Jao, Jarry1250, Jeandré du Toit, José Luís Ávila Silveira, Juliancolton Alternative, Kaisershatner, Kingvald, Kordas, Kpalion, Kransky, Krun, Latulla, LeonWhite, Lobsterthermidor, Lord Fitzwarin, Ludger1961, Lupo, László Szegedi, Mattis, Megan1967, Mich Taylor, Monsieur Net, Montrealais, Mr Gronk, MrDolomite, Narracan3824, Omnipaedista, OrangeDog, Oxymoron83, Pawel z Niepolomic, Petter Strandmark, PinchasC, QVanillaQ, Rbonvall, Reinyday, Rich Farmbrough, Robert1947, Schriste, Scott Martin, Scott Ritchie, ScottMainwaring, Septegram, Snori, Tamfang, Thehelpfulone, Tobyc75, Valentinian, Wilhelm meis, Yopie, ZanderSchubert, Zscout370, 44 anonymous edits

Saint-Aubin-Sauges *Source*: http://en.wikipedia.org/w/index.php?title=Saint-Aubin-Sauges *Contributors*: Adlei86, BillC, D6, DidiWeidmann, Docu, MadGeographer, Sole Soul, Tobyc75, 1 anonymous edits

Provence,_Switzerland *Source*: http://en.wikipedia.org/w/index.php?title=Provence%2C_Switzerland *Contributors*: Castrup.se, D6, Dirrival, Dl2000, Docu, Sole Soul, TCY, Tobyc75, Ultimate Destiny, 4 anonymous edits

Fresens *Source*: http://en.wikipedia.org/w/index.php?title=Fresens *Contributors*: Adlei86, BillC, D6, DidiWeidmann, Tamfang, Tobyc75, 1 anonymous edits

Image Sources, Licenses and Contributors

file:Montalchez-blason.png *Source*: http://en.wikipedia.org/w/index.php?title=File:Montalchez-blason.png *License*: unknown *Contributors*: Greatpatton, Manoillon, Voyager

file:Switzerland location map.svg *Source*: http://en.wikipedia.org/w/index.php?title=File:Switzerland_location_map.svg *License*: unknown *Contributors*: Lukasb1992, Maoman, NordNordWest, Sting, Wikisoft*, 1 anonymous edits

File:Red pog.svg *Source*: http://en.wikipedia.org/w/index.php?title=File:Red_pog.svg *License*: unknown *Contributors*: Anomie

File:Karte Gemeinden der Schweiz 2012.png *Source*: http://en.wikipedia.org/w/index.php?title=File:Karte_Gemeinden_der_Schweiz_2012.png *License*: unknown *Contributors*: Tschubby

File:Bevaix.gif *Source*: http://en.wikipedia.org/w/index.php?title=File:Bevaix.gif *License*: unknown *Contributors*: Auteur : Gérard Crameri Original uploader was Prog-x at fr.wikipedia

File:Flag of Boudry.svg *Source*: http://en.wikipedia.org/w/index.php?title=File:Flag_of_Boudry.svg *License*: unknown *Contributors*: Mathieu Despont

File:Brot-Dessous-coat of arms.svg *Source*: http://en.wikipedia.org/w/index.php?title=File:Brot-Dessous-coat_of_arms.svg *License*: unknown *Contributors*: User:Manoillon

File:Corcelles-Cormondrèche-coat of arms.svg *Source*: http://en.wikipedia.org/w/index.php?title=File:Corcelles-Cormondrèche-coat_of_arms.svg *License*: unknown *Contributors*: User:Manoillon

File:Cortaillod-blason.png *Source*: http://en.wikipedia.org/w/index.php?title=File:Cortaillod-blason.png *License*: unknown *Contributors*: Greatpatton, Manoillon, Voyager

File:Fresens-blason.png *Source*: http://en.wikipedia.org/w/index.php?title=File:Fresens-blason.png *License*: unknown *Contributors*: BillC, De728631, Greatpatton, Ludger1961, Manoillon, Voyager,

File:Gorgier-blason.png *Source*: http://en.wikipedia.org/w/index.php?title=File:Gorgier-blason.png *License*: unknown *Contributors*: Greatpatton, Manoillon, Voyager

File:Montalchez-blason.png *Source*: http://en.wikipedia.org/w/index.php?title=File:Montalchez-blason.png *License*: unknown *Contributors*: Greatpatton, Manoillon, Voyager

File:Peseux, Neuchâtel-coat of arms.svg *Source*: http://en.wikipedia.org/w/index.php?title=File:Peseux,_Neuchâtel-coat_of_arms.svg *License*: unknown *Contributors*: User:Manoillon

File:Rochefort-NE-blason.png *Source*: http://en.wikipedia.org/w/index.php?title=File:Rochefort-NE-blason.png *License*: unknown *Contributors*: Greatpatton, Manoillon, Perhelion, Skim, Voyager

File:Saint-Aubin-Sauges-coat of arms.svg *Source*: http://en.wikipedia.org/w/index.php?title=File:Saint-Aubin-Sauges-coat_of_arms.svg *License*: unknown *Contributors*: User:Manoillon

File:Vaumarcus-blason.png *Source*: http://en.wikipedia.org/w/index.php?title=File:Vaumarcus-blason.png *License*: unknown *Contributors*: BillC, Greatpatton, Manoillon, Voyager

File:Suisse cantons.svg *Source*: http://en.wikipedia.org/w/index.php?title=File:Suisse_cantons.svg *License*: unknown *Contributors*: User:Pymouss44

File:Karte 13 Alte Orte.png *Source*: http://en.wikipedia.org/w/index.php?title=File:Karte_13_Alte_Orte.png *License*: unknown *Contributors*: User:Sidonius

Image:Wappen Zürich matt.svg *Source*: http://en.wikipedia.org/w/index.php?title=File:Wappen_Zürich_matt.svg *License*: unknown *Contributors*: Sa-se

Image:Wappen Bern matt.svg *Source*: http://en.wikipedia.org/w/index.php?title=File:Wappen_Bern_matt.svg *License*: unknown *Contributors*: User:Sa-se

Image:Wappen Luzern matt.svg *Source*: http://en.wikipedia.org/w/index.php?title=File:Wappen_Luzern_matt.svg *License*: unknown *Contributors*: Sa-se

Image:Wappen Uri matt.svg *Source*: http://en.wikipedia.org/w/index.php?title=File:Wappen_Uri_matt.svg *License*: unknown *Contributors*: Sa-se

Image:Wappen Obwalden matt.svg *Source*: http://en.wikipedia.org/w/index.php?title=File:Wappen_Obwalden_matt.svg *License*: unknown *Contributors*: Heraldiker Rudolf Münger

Image:Wappen Nidwalden matt.svg *Source*: http://en.wikipedia.org/w/index.php?title=File:Wappen_Nidwalden_matt.svg *License*: unknown *Contributors*: Sa-se

Image:Wappen Glarus matt.svg *Source*: http://en.wikipedia.org/w/index.php?title=File:Wappen_Glarus_matt.svg *License*: unknown *Contributors*: Sa-se

Image:Wappen Zug matt.svg *Source*: http://en.wikipedia.org/w/index.php?title=File:Wappen_Zug_matt.svg *License*: unknown *Contributors*: Sa-se

Image:Wappen Freiburg matt.svg *Source*: http://en.wikipedia.org/w/index.php?title=File:Wappen_Freiburg_matt.svg *License*: unknown *Contributors*: Sa-se

Image:Wappen Solothurn matt.svg *Source*: http://en.wikipedia.org/w/index.php?title=File:Wappen_Solothurn_matt.svg *License*: unknown *Contributors*: Sa-se

Image:Wappen Basel-Stadt matt.svg *Source*: http://en.wikipedia.org/w/index.php?title=File:Wappen_Basel-Stadt_matt.svg *License*: unknown *Contributors*: Sa-se

Image:Coat of arms of Kanton Basel-Landschaft.svg *Source*: http://en.wikipedia.org/w/index.php?title=File:Coat_of_arms_of_Kanton_Basel-Landschaft.svg *License*: unknown *Contributors*: Wappenentwurf: Otto Plattner diese Datei: Jwnabd

Image:Wappen Schaffhausen matt.svg *Source*: http://en.wikipedia.org/w/index.php?title=File:Wappen_Schaffhausen_matt.svg *License*: unknown *Contributors*: (previous version maybe User:Delta-9)

Image:Wappen Appenzell Ausserrhoden matt.svg *Source*: http://en.wikipedia.org/w/index.php?title=File:Wappen_Appenzell_Ausserrhoden_matt.svg *License*: unknown *Contributors*: Sa-se

Image:Wappen Appenzell Innerrhoden matt.svg *Source*: http://en.wikipedia.org/w/index.php?title=File:Wappen_Appenzell_Innerrhoden_matt.svg *License*: unknown *Contributors*: Sa-se

Image:Coat of arms of canton of St. Gallen.svg *Source*: http://en.wikipedia.org/w/index.php?title=File:Coat_of_arms_of_canton_of_St._Gallen.svg *License*: unknown *Contributors*: Wappenentwurf: unbekannt diese Datei: Jwnabd

Image:Wappen Graubünden matt.svg *Source*: http://en.wikipedia.org/w/index.php?title=File:Wappen_Graubünden_matt.svg *License*: unknown *Contributors*: Sa-se

Image:Wappen Aargau matt.svg *Source*: http://en.wikipedia.org/w/index.php?title=File:Wappen_Aargau_matt.svg *License*: unknown *Contributors*: Sa-se

Image:Wappen Thurgau matt.svg *Source*: http://en.wikipedia.org/w/index.php?title=File:Wappen_Thurgau_matt.svg *License*: unknown *Contributors*: Sa-se

Image:Wappen Tessin matt.svg *Source*: http://en.wikipedia.org/w/index.php?title=File:Wappen_Tessin_matt.svg *License*: unknown *Contributors*: Abaddon1337, Ricordisamoa, Sa-se, Skim, 1 anonymous edits

Image:Wappen Waadt matt.svg *Source*: http://en.wikipedia.org/w/index.php?title=File:Wappen_Waadt_matt.svg *License*: unknown *Contributors*: Sa-se

Image:Wappen Wallis matt.svg *Source*: http://en.wikipedia.org/w/index.php?title=File:Wappen_Wallis_matt.svg *License*: unknown *Contributors*: Sa-se

Image:Wappen Neuenburg matt.svg *Source*: http://en.wikipedia.org/w/index.php?title=File:Wappen_Neuenburg_matt.svg *License*: unknown *Contributors*: Sa-se

Image:Wappen Genf matt.svg *Source*: http://en.wikipedia.org/w/index.php?title=File:Wappen_Genf_matt.svg *License*: unknown *Contributors*: Sa-se

Image:Wappen Jura matt.svg *Source*: http://en.wikipedia.org/w/index.php?title=File:Wappen_Jura_matt.svg *License*: unknown *Contributors*: Sa-se

Image:Coat of Arms of Switzerland.svg *Source*: http://en.wikipedia.org/w/index.php?title=File:Coat_of_Arms_of_Switzerland.svg *License*: unknown *Contributors*: User:E Pluribus Anthony, User:Reisio, User:Wiki-vr, User:Zscout370

Image:Karikatur Teilung Basels.jpg *Source*: http://en.wikipedia.org/w/index.php?title=File:Karikatur_Teilung_Basels.jpg *License*: unknown *Contributors*: Alarob, Aliman5040, Infrogmation, Mapmarks, Mattes, Parpan05, Roland zh, Sidonius, Stefan Kühn

File:Blason Ducs Bourgogne (ancien).svg *Source*: http://en.wikipedia.org/w/index.php?title=File:Blason_Ducs_Bourgogne_(ancien).svg *License*: unknown *Contributors*: User:Zigeuner

File:Flag of Switzerland.svg *Source*: http://en.wikipedia.org/w/index.php?title=File:Flag_of_Switzerland.svg *License*: unknown *Contributors*: User:-xfi-, User:Marc Mongenet, User:Zscout370

File:Wappen Neuenburg.svg *Source*: http://en.wikipedia.org/w/index.php?title=File:Wappen_Neuenburg.svg *License*: unknown *Contributors*: User:Sidonius

File:Le Locle 1907.JPG *Source*: http://en.wikipedia.org/w/index.php?title=File:Le_Locle_1907.JPG *License*: unknown *Contributors*: Original uploader was Steffen M. at de.wikipedia

File:Lac de Neuchatel.jpg *Source*: http://en.wikipedia.org/w/index.php?title=File:Lac_de_Neuchatel.jpg *License*: unknown *Contributors*: Jean-Claude Collaud

File:Château de Neuchâtel.jpg *Source*: http://en.wikipedia.org/w/index.php?title=File:Château_de_Neuchâtel.jpg *License*: unknown *Contributors*: Calendula, FSII, Kurpfalzbilder.de

File:Karte Kanton Neuenburg Bezirke.png *Source*: http://en.wikipedia.org/w/index.php?title=File:Karte_Kanton_Neuenburg_Bezirke.png *License*: unknown *Contributors*: Juiced lemon, Tschubby

File:La Chaux de Fonds.jpg *Source*: http://en.wikipedia.org/w/index.php?title=File:La_Chaux_de_Fonds.jpg *License*: unknown *Contributors*: User:MadGeographer

Image:Elementy herbu.jpg *Source*: http://en.wikipedia.org/w/index.php?title=File:Elementy_herbu.jpg *License*: unknown *Contributors*: User:Qyd, User:Steifer

File:Hyghalmen Roll Late 1400s.jpg *Source*: http://en.wikipedia.org/w/index.php?title=File:Hyghalmen_Roll_Late_1400s.jpg *License*: unknown *Contributors*: AndreasPraefcke, Chatsam, Evadb, G.dallorto, Jaybear, Knorrepoes, LeoDavid, Massimop, Para, Rosenzweig, Skim, Technokrat, 1 anonymous edits

File:University of Plymouth Coat of Arms.png *Source*: http://en.wikipedia.org/w/index.php?title=File:University_of_Plymouth_Coat_of_Arms.png *License*: unknown *Contributors*: Garter, Clarenceux and Norroy and Ulster King of Arms.

File:Vaasa.vaakuna.svg *Source*: http://en.wikipedia.org/w/index.php?title=File:Vaasa.vaakuna.svg *License*: unknown *Contributors*: User:Care

File:Ottoman flag.svg *Source*: http://en.wikipedia.org/w/index.php?title=File:Ottoman_flag.svg *License*: unknown *Contributors*: AnonMoos, Cycn, David0811, FSII, Homo lupus, Quest for Truth, Ricordisamoa, TFCforever, TRAJAN 117, Mo.лox, 1 anonymous edits

Image:Coats of arms of the Kingdom of Egypt and Sudan.png *Source*: http://en.wikipedia.org/w/index.php?title=File:Coats_of_arms_of_the_Kingdom_of_Egypt_and_Sudan.png *License*: unknown *Contributors*: User:Beao, User:CrimeCentral, User:SreeBot

Image:Coat of arms of United Arab Republic (Syria 1958-61, Egypt 1958-1971).svg *Source*: http://en.wikipedia.org/w/index.php?title=File:Coat_of_arms_of_United_Arab_Republic_(Syria_1958-61,_Egypt_1958-1971).svg *License*: unknown *Contributors*: User:AnonMoos, User:Kafka1, User:Omar86, User:R-41, User:Tonyjeff

File:Mitsubishi_logo.svg *Source*: http://en.wikipedia.org/w/index.php?title=File:Mitsubishi_logo.svg *License*: unknown *Contributors*: User:Monaneko

File:John paul 2 coa.svg *Source*: http://en.wikipedia.org/w/index.php?title=File:John_paul_2_coa.svg *License*: unknown *Contributors*: User:mAgul

File:PD-icon.svg *Source*: http://en.wikipedia.org/w/index.php?title=File:PD-icon.svg *License*: unknown *Contributors*: Alex.muller, Anomie, Anonymous Dissident, CBM, MBisanz, PBS, Quadell, Rocket000, Strangerer, Timotheus Canens, 1 anonymous edits

File:Azure, a bend Or.svg *Source*: http://en.wikipedia.org/w/index.php?title=File:Azure,_a_bend_Or.svg *License*: unknown *Contributors*: User:Bdesham, User:Halibutt

File:Wappen Behnsdorf.png *Source*: http://en.wikipedia.org/w/index.php?title=File:Wappen_Behnsdorf.png *License*: unknown *Contributors*: Jörg Mantzsch

File:Quarterly demo.svg *Source*: http://en.wikipedia.org/w/index.php?title=File:Quarterly_demo.svg *License*: unknown *Contributors*: User:Balmung0731

File:Brandenburg Wappen.svg *Source*: http://en.wikipedia.org/w/index.php?title=File:Brandenburg_Wappen.svg *License*: unknown *Contributors*: User:David Liuzzo

File:COA fr BRE.svg *Source*: http://en.wikipedia.org/w/index.php?title=File:COA_fr_BRE.svg *License*: unknown *Contributors*: User:Brieg

File:PB Ostergotland CoA.png *Source*: http://en.wikipedia.org/w/index.php?title=File:PB_Ostergotland_CoA.png *License*: unknown *Contributors*: AtonX, BenAllard, GeorgHH, Halibutt, Kilom691, Liftarn, Lokal Profil, 1 anonymous edits

File:Coa Hungary Country History Mid (1867).svg *Source*: http://en.wikipedia.org/w/index.php?title=File:Coa_Hungary_Country_History_Mid_(1867).svg *License*: unknown *Contributors*: User:Madboy74

file:Saint-Aubin-Sauges-coat of arms.svg *Source*: http://en.wikipedia.org/w/index.php?title=File:Saint-Aubin-Sauges-coat_of_arms.svg *License*: unknown *Contributors*: User:Manoillon

File:St-Aubin-Sauges Port Conty DSC9855 20110923.jpg *Source*: http://en.wikipedia.org/w/index.php?title=File:St-Aubin-Sauges_Port_Conty_DSC9855_20110923.jpg *License*: unknown *Contributors*: User:Imagepro

File:La Béroche établissement gallo-romain.jpg *Source*: http://en.wikipedia.org/w/index.php?title=File:La_Béroche_établissement_gallo-romain.jpg *License*: unknown *Contributors*: Imagepro - Denis Beyeler 2011

File:St-Aubin-Sauges Port Conty DSC9853 20110923.jpg *Source*: http://en.wikipedia.org/w/index.php?title=File:St-Aubin-Sauges_Port_Conty_DSC9853_20110923.jpg *License*: unknown *Contributors*: User:Imagepro

file:Provence.jpg *Source*: http://en.wikipedia.org/w/index.php?title=File:Provence.jpg *License*: unknown *Contributors*: unbekannt. Original uploader was Ambroix at de.wikipedia

file:Fresens-blason.png *Source*: http://en.wikipedia.org/w/index.php?title=File:Fresens-blason.png *License*: unknown *Contributors*: BillC, De728631, Greatpatton, Ludger1961, Manoillon, Voyager,

GNU Free Documentation License Version 1.2, November 2002 Copyright (C) 2000,2001,2002 Free Software Foundation, Inc. 59 Temple Place, Suite 330, Boston, MA 02111-1307 USA Everyone is permitted to copy and distribute verbatim copies of this license document, but changing it is not allowed.

0. PREAMBLE

The purpose of this License is to make a manual, textbook, or other functional and useful document "free" in the sense of freedom: to assure everyone the effective freedom to copy and redistribute it, with or without modifying it, either commercially or noncommercially. Secondarily, this License preserves for the author and publisher a way to get credit for their work, while not being considered responsible for modifications made by others. This License is a kind of "copyleft", which means that derivative works of the document must themselves be free in the same sense. It complements the GNU General Public License, which is a copyleft license designed for free software. We have designed this License in order to use it for manuals for free software, because free software needs free documentation: a free program should come with manuals providing the same freedoms that the software does. But this License is not limited to software manuals; it can be used for any textual work, regardless of subject matter or whether it is published as a printed book. We recommend this License principally for works whose purpose is instruction or reference.

1. APPLICABILITY AND DEFINITIONS

This License applies to any manual or other work, in any medium, that contains a notice placed by the copyright holder saying it can be distributed under the terms of this License. Such a notice grants a world-wide, royalty-free license, unlimited in duration, to use that work under the conditions stated herein. The "Document", below, refers to any such manual or work. Any member of the public is a licensee, and is addressed as "you". You accept the license if you copy, modify or distribute the work in a way requiring permission under copyright law. A "Modified Version" of the Document means any work containing the Document or a portion of it, either copied verbatim, or with modifications and/or translated into another language. A "Secondary Section" is a named appendix or a front-matter section of the Document that deals exclusively with the relationship of the publishers or authors of the Document to the Document's overall subject (or to related matters) and contains nothing that could fall directly within that overall subject. (Thus, if the Document is in part a textbook of mathematics, a Secondary Section may not explain any mathematics.) The relationship could be a matter of historical connection with the subject or with related matters, or of legal, commercial, philosophical, ethical or political position regarding them. The "Invariant Sections" are certain Secondary Sections whose titles are designated, as being those of Invariant Sections, in the notice that says that the Document is released under this License. If a section does not fit the above definition of Secondary then it is not allowed to be designated as Invariant. The Document may contain zero Invariant Sections. If the Document does not identify any Invariant Sections then there are none. The "Cover Texts" are certain short passages of text that are listed, as Front-Cover Texts or Back-Cover Texts, in the notice that says that the Document is released under this License. A Front-Cover Text may be at most 5 words, and a Back-Cover Text may be at most 25 words. A "Transparent" copy of the Document means a machine-readable copy, represented in a format whose specification is available to the general public, that is suitable for revising the document straightforwardly with generic text editors or (for images composed of pixels) generic paint programs or (for drawings) some widely available drawing editor, and that is suitable for input to text formatters or for automatic translation to a variety of formats suitable for input to text formatters. A copy made in an otherwise Transparent file format whose markup, or absence of markup, has been arranged to thwart or discourage subsequent modification by readers is not Transparent. An image format is not Transparent if used for any substantial amount of text. A copy that is not "Transparent" is called "Opaque". Examples of suitable formats for Transparent copies include plain ASCII without markup, Texinfo input format, LaTeX input format, SGML or XML using a publicly available DTD, and standard-conforming simple HTML, PostScript or PDF designed for human modification. Examples of transparent image formats include PNG, XCF and JPG. Opaque formats include proprietary formats that can be read and edited only by proprietary word processors, SGML or XML for which the DTD and/or processing tools are not generally available, and the machine-generated HTML, PostScript or PDF produced by some word processors for output purposes only. The "Title Page" means, for a printed book, the title page itself, plus such following pages as are needed to hold, legibly, the material this License requires to appear in the title page. For works in formats which do not have any title page as such, "Title Page" means the text near the most prominent appearance of the work's title, preceding the beginning of the body of the text. A section "Entitled XYZ" means a named subunit of the Document whose title either is precisely XYZ or contains XYZ in parentheses following text that translates XYZ in another language. (Here XYZ stands for a specific section name mentioned below, such as "Acknowledgements", "Dedications", "Endorsements", or "History".) To "Preserve the Title" of such a section when you modify the Document means that it remains a section "Entitled XYZ" according to this definition. The Document may include Warranty Disclaimers next to the notice which states that this License applies to the Document. These Warranty Disclaimers are considered to be included by reference in this License, but only as regards disclaiming warranties: any other implication that these Warranty Disclaimers may have is void and has no effect on the meaning of this License.

2. VERBATIM COPYING

You may copy and distribute the Document in any medium, either commercially or noncommercially, provided that this License, the copyright notices, and the license notice saying this License applies to the Document are reproduced in all copies, and that you add no other conditions whatsoever to those of this License. You may not use technical measures to obstruct or control the reading or further copying of the copies you make or distribute. However, you may accept compensation in exchange for copies. If you distribute a large enough number of copies you must also follow the conditions in section 3. You may also lend copies, under the same conditions stated above, and you may publicly display copies.

3. COPYING IN QUANTITY

If you publish printed copies (or copies in media that commonly have printed covers) of the Document, numbering more than 100, and the Document's license notice requires Cover Texts, you must enclose the copies in covers that carry, clearly and legibly, all these Cover Texts: Front-Cover Texts on the front cover, and Back-Cover Texts on the back cover. Both covers must also clearly and legibly identify you as the publisher of these copies. The front cover must present the full title with all words of the title equally prominent and visible. You may add other material on the covers in addition. Copying with changes limited to the covers, as long as they preserve the title of the Document and satisfy these conditions, can be treated as verbatim copying in other respects. If the required texts for either cover are too voluminous to fit legibly, you should put the first ones listed (as many as fit reasonably) on the actual cover, and continue the rest onto adjacent pages. If you publish or distribute Opaque copies of the Document numbering more than 100, you must either include a machine-readable Transparent copy along with each Opaque copy, or state in or with each Opaque copy a computer-network location from which the general network-using public has access to download using public-standard network protocols a complete Transparent copy of the Document, free of added material. If you use the latter option, you must take reasonably prudent steps, when you begin distribution of Opaque copies in quantity, to ensure that this Transparent copy will remain thus accessible at the stated location until at least one year after the last time you distribute an Opaque copy (directly or through your agents or retailers) of that edition to the public. It is requested, but not required, that you contact the authors of the Document well before redistributing any large number of copies, to give them a chance to provide you with an updated version of the Document.

4. MODIFICATIONS

You may copy and distribute a Modified Version of the Document under the conditions of sections 2 and 3 above, provided that you release the Modified Version under precisely this License, with the Modified Version filling the role of the Document, thus licensing distribution and modification of the Modified Version to whoever possesses a copy of it. In addition, you must do these things in the Modified Version: A. Use in the Title Page (and on the covers, if any) a title distinct from that of the Document, and from those of previous versions (which should, if there were any, be listed in the History section of the Document). You may use the same title as a previous version if the original publisher of that version gives permission. B. List on the Title Page, as authors, one or more persons or entities responsible for authorship of the modifications in the Modified Version, together with at least five of the principal authors of the Document (all of its principal authors, if it has fewer than five), unless they release you from this requirement. C. State on the Title page the name of the publisher of the Modified Version, as the publisher. D. Preserve all the copyright notices of the Document. E. Add an appropriate copyright notice for your modifications adjacent to the other copyright notices. F. Include, immediately after the copyright notices, a license notice giving the public permission to use the Modified Version under the terms of this License, in the form shown in the Addendum below. G. Preserve in that license notice the full lists of Invariant Sections and required Cover Texts given in the Document's license notice. H. Include an unaltered copy of this License. I. Preserve the section Entitled "History", Preserve its Title, and add to it an item stating at least the title, year, new authors, and publisher of the Modified Version as given on the Title Page. If there is no section Entitled "History" in the Document, create one stating the title, year, authors, and publisher of the Document as given on its Title Page, then add an item describing the Modified Version as stated in the previous sentence. J. Preserve the network location, if any, given in the Document for public access to a Transparent copy of the Document, and likewise the network locations given in the Document for previous versions it was based on. These may be placed in the "History" section. You may omit a network location for a work that was published at least four years before the Document itself, or if the original publisher of the version it refers to gives permission. K. For any section Entitled "Acknowledgements" or "Dedications", Preserve the Title of the section, and preserve in the section all the substance and tone of each of the contributor acknowledgements and/or dedications given therein. L. Preserve all the Invariant Sections of the Document, unaltered in their text and in their titles. Section numbers or the equivalent are not considered part of the section titles. M. Delete any section Entitled "Endorsements". Such a section may not be included in the Modified Version. N. Do not retitle any existing section to be Entitled "Endorsements" or to conflict in title with any Invariant Section. O. Preserve any Warranty Disclaimers. If the Modified Version includes new front-matter sections or appendices that qualify as Secondary Sections and contain no material copied from the Document, you may at your option designate some or all of these sections as invariant. To do this, add their titles to the list of Invariant Sections in the Modified Version's license notice. These titles must be distinct from any other section titles. You may add a section Entitled "Endorsements", provided it contains nothing but endorsements of your Modified Version by various parties--for example, statements of peer review or that the text has been approved by an organization as the authoritative definition of a standard. You may add a passage of up to five words as a Front-Cover Text, and a passage of up to 25 words as a Back-Cover Text, to the end of the list of Cover Texts in the Modified Version. Only one passage of Front-Cover Text and one of Back-Cover Text may be added by (or through arrangements made by) any one entity. If the Document already includes a cover text for the same cover, previously added by you or by arrangement made by the same entity you are acting on behalf of, you may not add another; but you may replace the old one, on explicit permission from the previous publisher that added the old one. The author(s) and publisher(s) of the Document do not by this License give permission to use their names for publicity for or to assert or imply endorsement of any Modified Version.

5. COMBINING DOCUMENTS

You may combine the Document with other documents released under this License, under the terms defined in section 4 above for modified versions, provided that you include in the combination all of the Invariant Sections of all of the original documents, unmodified, and list them all as Invariant Sections of your combined work in its license notice, and that you preserve all their Warranty Disclaimers. The combined work need only contain one copy of this License, and multiple identical Invariant Sections may be replaced with a single copy. If there are multiple Invariant Sections with the same name but different contents, make the title of each such section unique by adding at the end of it, in parentheses, the name of the original author or publisher of that section if known, or else a unique number. Make the same adjustment to the section titles in the list of Invariant Sections in the license notice of the combined work. In the combination, you must combine any sections Entitled "History" in the various original documents, forming one section Entitled "History"; likewise combine any sections Entitled "Acknowledgements", and any sections Entitled "Dedications". You must delete all sections Entitled "Endorsements".

6. COLLECTIONS OF DOCUMENTS

You may make a collection consisting of the Document and other documents released under this License, and replace the individual copies of this License in the various documents with a single copy that is included in the collection, provided that you follow the rules of this License for verbatim copying of each of the documents in all other respects. You may extract a single document from such a collection, and distribute it individually under this License, provided you insert a copy of this License into the extracted document, and follow this License in all other respects regarding verbatim copying of that document.

7. AGGREGATION WITH INDEPENDENT WORKS

A compilation of the Document or its derivatives with other separate and independent documents or works, in or on a volume of a storage or distribution medium, is called an "aggregate" if the copyright resulting from the compilation is not used to limit the legal rights of the compilation's users beyond what the individual works permit. When the Document is included in an aggregate, this License does not apply to the other works in the aggregate which are not themselves derivative works of the Document. If the Cover Text requirement of section 3 is applicable to these copies of the Document, then if the Document is less than one half of the entire aggregate, the Document's Cover Texts may be placed on covers that bracket the Document within the aggregate, or the electronic equivalent of covers if the Document is in electronic form. Otherwise they must appear on printed covers that bracket the whole aggregate.

8. TRANSLATION

Translation is considered a kind of modification, so you may distribute translations of the Document under the terms of section 4. Replacing Invariant Sections with translations requires special permission from their copyright holders, but you may include translations of some or all Invariant Sections in addition to the original versions of these Invariant Sections. You may include a translation of this License, and all the license notices in the Document, and any Warranty Disclaimers, provided that you also include the original English version of this License and the original versions of those notices and disclaimers. In case of a disagreement between the translation and the original version of this License or a notice or disclaimer, the original version will prevail. If a section in the Document is Entitled "Acknowledgements", "Dedications", or "History", the requirement (section 4) to Preserve its Title (section 1) will typically require changing the actual title.

9. TERMINATION

You may not copy, modify, sublicense, or distribute the Document except as expressly provided for under this License. Any other attempt to copy, modify, sublicense or distribute the Document is void, and will automatically terminate your rights under this License. However, parties who have received copies, or rights, from you under this License will not have their licenses terminated so long as such parties remain in full compliance.

10. FUTURE REVISIONS OF THIS LICENSE

The Free Software Foundation may publish new, revised versions of the GNU Free Documentation License from time to time. Such new versions will be similar in spirit to the present version, but may differ in detail to address new problems or concerns. See http://www.gnu.org/copyleft/. Each version of the License is given a distinguishing version number. If the Document specifies that a particular numbered version of this License "or any later version" applies to it, you have the option of following the terms and conditions either of that specified version or of any later version that has been published (not as a draft) by the Free Software Foundation. If the Document does not specify a version number of this License, you may choose any version ever published (not as a draft) by the Free Software Foundation. ADDENDUM: How to use this License for your documents To use this License in a document you have written, include a copy of the License in the document and put the following copyright and license notices just after the title page: Copyright (c) YEAR YOUR NAME. Permission is granted to copy, distribute and/or modify this document under the terms of the GNU Free Documentation License, Version 1.2 or any later version published by the Free Software Foundation; with no Invariant Sections, no Front-Cover Texts, and no Back-Cover Texts. A copy of the license is included in the section entitled "GNU Free Documentation License". If you have Invariant Sections, Front-Cover Texts and Back-Cover Texts, replace the "with...Texts." line with this: with the Invariant Sections being LIST THEIR TITLES, with the Front-Cover Texts being LIST, and with the Back-Cover Texts being LIST. If you have Invariant Sections without Cover Texts, or some other combination of the three, merge those two alternatives to suit the situation. If your document contains nontrivial examples of program code, we recommend releasing these examples in parallel under your choice of free software license, such as the GNU General Public License, to permit their use in free software.

MIX
Papier aus verantwortungsvollen Quellen
Paper from responsible sources
FSC® C105338

Printed by Books on Demand GmbH, Norderstedt / Germany